Dynamics of
Brain Monoamines

Dynamics of
Brain Monoamines

J. C. de la Torre

Division of Neurological Surgery
The University of Chicago
Pritzker School of Medicine
Chicago, Illinois

℗ **PLENUM PRESS • NEW YORK-LONDON • 1972**

Library of Congress Catalog Card Number 75-183560
ISBN-13: 978-1-4684-1946-7   e-ISBN-13: 978-1-4684-1944-3
DOI: 10.1007/978-1-4684-1944-3

© 1972 Plenum Press, New York
Softcover reprint of the hardcover 1st edition 1972
A Division of Plenum Publishing Corporation
227 West 17th Street, New York, N.Y. 10011
United Kingdom edition published by Plenum Press, London
A Division of Plenum Publishing Company, Ltd.
Davis House (4th Floor), 8 Scrubs Lane, Harlesden, London, NW10 6SE, England

To
Santiago Ramón y Cajal,
who opened our eyes to the intricate beauty of
the nervous system

Preface

The noted French physiologist Claude Bernard was fond of noting that "experimental ideas are often born by chance, with the help of some casual observation."

If history teaches us, we realize that this statement is as true today as it was 100 years ago. Today, however, we are luckier. We have, generally, some basis from previous work so that we can follow a line of research and thus provide even our most fundamental studies with an air of scientific respectability. Present-day research is rather like the working of a giant crossword puzzle where some of the letters or even words have been completed and the approach toward a relatively confident solution is a matter of juggling a series of permutations.

I often wonder, though, how it was for the early brain researchers, who had very few letters, and, even less, no words, completed in their research puzzles to augment intelligent casual observation. By "early," I refer to nineteenth or early twentieth century investigators—men like Cajal, Sherrington, Pavlov, and Sechenov, whose investigations did more to advance their field in a span of 50 years than all of the cumulative previous work done on the brain and nervous system since the ancient Greeks.

Although many books had been written prior to the works of Sherrington, Cajal, and the others, they were in many ways useless and at times a hindrance to research. Whatever inspiration possessed the early great investigators to make the first connections between complex neurobiological phenomena, their accomplishments do mirror those of other men, in other periods of time, who seemed also inspired in creative work—men like Bernini, Renoir, Dostoevsky, and Newton, who revolutionized their fields for others to follow.

Consider, for example, that it took man about a million years to learn the fine art of flint chipping. One can almost envision the neolithic man pounding, chewing, and tearing up everything in his environment, not unlike a modern young child or an infant gorilla. It is this capacity in man to explore and be curious that has led him to transcend his biological role of reproduction and simple survival.

It is certainly not immodest or whimsical to say that the study of the brain and its functions is one of the most difficult tasks a man can set for himself. It is in this sense, then, that I for one feel very grateful for the many people who have paved the way with their knowledge and results so that others, myself included, can pursue an idea or a project on a more certain footing.

The field of monoamine research is changing so rapidly and advancing so fast that some of the theoretical expectations discussed in this volume may become reality even as the experimental evidence is presented. The neurochemical and pharmacological study of the brain is an area of electrifying potential, Orwell's *1984* "phantastica" drugs come sooner. Within 50 to 80 years we may discover a series of psychoactive drugs that will modulate or induce certain behavioral reactions, such as control violent aggression, provide short- or long-term hibernation, prolong, shorten, or eradicate pleasant and unpleasant memory, stimulate high sexual responses, and heighten creative talent and learning ability. Perhaps later we may find the drug we are all seeking—one that induces the feeling of contentment in spite of circumstances. This and other aspects of neuropharmacology raise many questions of philosophy, ethics, religion, and economy.

Yet the mere fact that work has already begun in unraveling the mystery of the mind and, consequently, in directing our first steps toward discovering what we are justifies, in my opinion, further and more probing research in this field. Man has thus far been able to adapt to scientific progress, and although in some instances it may require one step backward in order to move two steps forward, these biological predictions are invariably inevitable.

I wish to express my appreciation to our two excellent medical secretaries who typed and retyped this manuscript, Mrs. Emma Balchunas and Mrs. Alice Davis. My many thanks to our medical artists, Miss Josephine Wu and Mr. Charles S. Wellek, who are responsible for many of the illustrations in the text; to our Photography Department and Mr. Libéro Giannini; and in particular to Dr. John F. Mullan, Acting Chairman of the Department of Surgery and Chief of Neurological Surgery, whose constant interest and encouragement in experimental research make it a pleasure to be associated with a clinical department.

J.C. de la Torre
November 1971

Introduction

Since the discovery of monoamines some 24 years ago by von Euler, many theories have been advanced to clarify their functional role in the central nervous system. What makes the monoamines (dopamine, noradrenaline, adrenaline, and serotonin) a subject of intensive interest and research is their possible role in brain function, especially in central synaptic transmission.

Interest in the field of monoamines received a boost with the awarding of the 1970 Nobel Prize in Medicine and Physiology to three men who laid the groundwork for much of the present research in neurotransmitter function: Julius Axelrod of the United States, Bernard Katz of Great Britain, and Ulf von Euler of Sweden.

It is becoming more and more evident that normal brain neuronal function may be dependent upon a metabolic equilibrium of central-acting amines. A disruption of this metabolic balance might result in various pathological states. The work of Maynert and Levi (1964), for example, has implicated brain stem activity of noradrenaline in various stress situations in the rat. Rage reactions and mood may be modified by electrical stimulation of the amygdala and septal regions, while various drugs modifying monoamine concentrations in the brain may induce depression or excitation in the animal. Monoamines have also been implicated in the regulation of other functions vital to the organism such as sleep, brain barrier mechanisms, temperature control, and circadian rhythms.

Problems relating to these states are usually studied by investigators trained in many different fields. Whether it is the pharmacologist, biochemist, anatomist, psychologist, physiologist, pathologist, psychiatrist, or any other in an allied field, there comes a time when communication of scientific data is hampered by the inability to share this knowledge through a common jargon, one that can be understood by most people working in the monoamine field. Interdisciplinary stumbling blocks have been the source of much misunderstanding in evaluating the role of monoamines in the brain. Besides the misunderstandings, there is always the strong possibility that a finding may never be seen by the busy investigator, who may have no more time than to scan one or two abstract journals. It is this problem that makes a good bibliographical report on the monoamines seem somewhat selective or rather lacking in content. One would hope that a journal devoted to reporting laboratory and clinical findings about the monoamines might become available. Such a journal would not be the answer to various minor details but should certainly bridge the inter-disciplinary gap in many fields.

The wide and uneven subcellular distribution of monoamines in living organisms has made them the subject of much speculation. It should be noted, however, that even though the bulk of research on their metabolic activity indicates that they function as neurotransmitters, final evidence is still lacking. Since monoamines are contained in neuronal tissue, it becomes at times tempting to infer that biological activity originates in or is related to amine changes. As a consequence, some interesting theories and some rather farfetched interpretations have proliferated.

There are many possible methods of elucidating amine function. Among the most promising are some discussed in subsequent chapters. These approaches include neuropharmacological, ultrastructural, auto-radiographic, biochemical, surgical, and histochemical fluorescence techniques. Histochemical fluorescence is the most recent research tool and perhaps the most difficult to interpret. It is at the same time a highly specific and versatile technique that has done more to advance our knowledge of monoamines in a few years than have most other methods that have been available for a longer time.

The idea that function of monoamines may be classified according to their action on sympathetic or parasympathetic systems is of considerable interest. It was first suggested by Brodie, based on the classical experiments of W. R. Hess on brain electrical stimulation. The theory states that catecholamines may mediate ergotropic zones (sympathetic) while serotonin mediates trophotropic zones (parasympathetic). Were one to admit such a premise, classification of psychoactive agents would depend on their action in the living system. Amphetamines, for example, would be considered ergotropic, while benzoquinolizine derivatives and reserpine would be trophotropic. Following this line of reasoning, LSD-25 would be either tropholytic or ergotropic and the phenothiazines would be ergolytic. However, drug action and amine changes may not always fall into a pat category, and even cautious interpretation of their functional consequence on each system may not always be that realistic.

The reason for the wide interest in psychotomimetic drugs is that, just as their name implies, they mimic various psychotic reactions, among schizophrenia. By studying the action of these compounds, it is hoped that information regarding therapy and etiology of some mental disorders may be possible. Dopaminergic pathways in nigrostriatal structures point to the importance of this amine in extrapyramidal disorders, notably Parkinson's disease. This also correlates with recent evidence of an enzymatic blood–brain barrier mechanism for the dopamine precursor L-dopa. When L-dopa is administered following enzymatic inhibition of peripheral decarboxylase activity, motor function and extrapyramidal symptoms show marked improvement. Monoamine-containing neurons affecting sleep mechanisms suggest a role for noradrenaline in paradoxical or rapid eye movement sleep, while serotonin would appear to be involved in slow wave sleep activity. Genetic and psychopathological aspects have further revealed the importance of understanding the molecular events resulting from monoamine imbalance in order to understand behavior.

Investigation of the role of noradrenaline in the mediation of vascular mechanisms, especially the cerebrovascularity, has opened a new field of possibilities in the understanding and treatment of traumatic head injury. As more data are collected, other substances such as dopamine are also being linked to the control of vasotonic mechanics, and it is at this level that knowledge concerning the blood–brain barrier permeability to the monoamine precursors becomes particularly fruitful.

This presentation is offered not so much to review old literature on the subject but, more hopefully, to present some stimulating aspects of interdisciplinary monoamine research. For such purpose, recent and original experimental evidence collectively linking the monoamines to various metabolic schemes will be discussed.

Contents

Chapter 1

Metabolism of Monoamines

1.1. BIOCHEMISTRY OF CATECHOLAMINES IN BRAIN

The term *catecholamine* refers to a group of substances with low molecular weight that contain a catechol nucleus (i.e., two hydroxyl ions, on the 3- and 4-positions) and an amine on the side chain (Fig. 1). The numbering system begins with the carbon adjacent to the side chain (with the first carbon atom called *alpha,* the second *beta,* etc.) and rotates counterclockwise within the phenol ring. *Monoamine* is an ambiguous term that applies to any aliphatic, aromatic, or heterocyclic compound in which a hydrogen atom has been replaced by an amine group. In the field of cerebral research, however, the term has been used to refer specifically to four substances found in the brain and the rest of the central nervous system: two primary amines, dopamine and noradrenaline; one secondary amine, adrenaline; and an indolealkylamine, serotonin. For the purpose of standardization, *monoamine* in this text will refer only to these four compounds.

The sequence of reactions for the conversion of adrenaline from dopamine and other amine precursors seems today fairly well established (see Fig. 2). Work by Blaschko (1939) first showed that the enzyme dopa decarboxylase is the catalyst in one of the reactions that leads from tyrosine to adrenaline. Gurin and Delluva (1947) confirmed this finding when they administered labeled phenylalanine to rats and recovered radioactive adrenaline from the adrenal glands. Experimental work also compared the

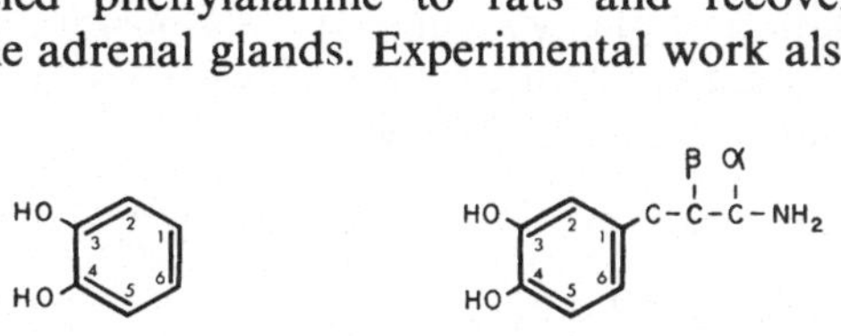

CATECHOL NUCLEUS   CATECHOLAMINE

Fig. 1

1

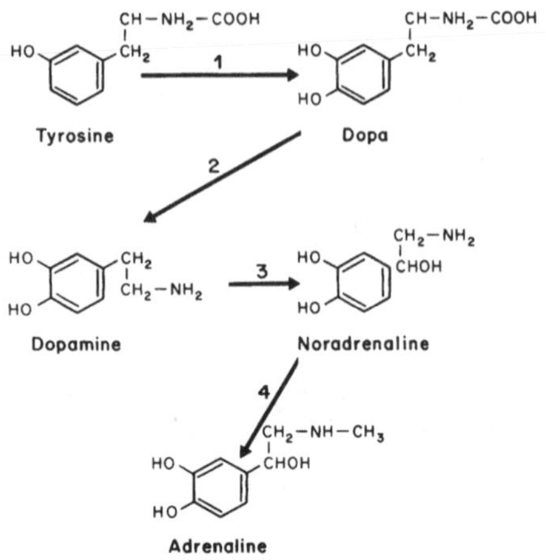

Fig. 2. Metabolism of noradrenaline and adrenaline in mammalian brain. (1) tyrosine hydroxylase; (2) dopa decarboxylase; (3) dopamine-β-hydroxylase; (4) phenylethanol-amine-N-methyltransferase.

phenolic hydroxyl groups in dopa decarboxylase and the bacterial enzyme tyrosine decarboxylase. The difference is that in the bacterial enzyme the hydroxyl group in position *para* is important for rapid decarboxylation, while in the mammalian enzyme the group involved is in the *meta* position to the side chain. The importance of the position of the hydroxyl group is shown by the fact that *meta*tyrosine and *ortho*tyrosine are decarboxylated by dopa decarboxylase, while the bacterial enzyme decarboxylates tyrosine and *meta*tyrosine but not *ortho*tyrosine. The nonspecificity of dopa decarb-oxylase has been shown by Lovenberg *et al.* (1962), who have demonstrated that this enzyme can decarboxylate a number of amino acids including phenylalanine, histidine, and the precursors of serotonin, tryptophan and 5-hydroxytryptophan.

1.1.1. Conversion of Phenylalanine to Tyrosine

A hydroxylation of phenylalanine at the 4-position occurs through the influence of phenylalanine hydroxylase, a liver enzyme. This conversion is brought about in two stages as follows:

Phenylalanine tetrahydrofolic acid $+ O_2 \xrightarrow{\text{tyrosine}}$ dihydrofolic acid

Dihydrofolic acid $+ NADPH_2 \xrightarrow{\text{enzyme}} NADP +$ tetrahydrofolic acid

This conversion is not indispensable because exogenous tyrosine can take the role of precursor in the second stage.

1.1.2. Conversion of Tyrosine to L-Dopa

The synthesis of catecholamines begins with the uptake of tyrosine from the circulation by various organs. Hydroxylation of tyrosine occurs at the 3-position through L-tyrosine hydroxylase, an enzyme that is found in all cells capable of synthesizing the catecholamines and is the rate-limiting step in their synthesis. The intracellular localization of tyrosine hydroxylase is still not known. After hydroxylation of tyrosine, L-dihydroxy-phenylalanine (L-dopa) is formed.

1.1.3. Conversion of L-Dopa to Dopamine

Aromatic L-amino acid decarboxylase or dopa decarboxylase, found as a nonspecific enzyme in the adrenal medulla, adrenergic endings, and more recently in brain capillaries (see Section 3.6), is responsible for the conversion of L-dopa to dopamine. This conversion is probably very rapid since practically no L-dopa can be detected in blood or the brain. A cofactor, pyridoral phosphate, is required also to form a complex that is slowly decarboxylated. Substances that slow down or block the formation of this complex will accelerate the synthesis to dopamine. The nonspecificity of dopa decarboxylase explains why this substance can decarboxylate α-aromatic and heterocyclic amino acids as well as their α-methylated acid derivatives, aromatic metaphenolic α-amines, and in particular α-methyl-*m*-tyrosine and α-methyldopa.

Dopamine, the first catecholamine synthesized by the organism, is found in mammalian brain. The importance of dopamine in motor functions is evident from three facts: (a) large amounts of dopamine are present in the neostriatum, which forms an important part of the extrapyramidal system; (b) extrapyramidal actions of reserpine deplete dopamine from the neostriatum; and (c) L-dopa is capable of counteracting the hypokinetic effects produced by reserpine in animals, a phenomenon that may be due to the formation of endogenous dopamine. Dopamine has been found in a wide variety of mammalian brains and is of particular interest because of its possible role in the mediation of dopaminergic pathways in cerebral structures such as the neostriatum, the substantia nigra, and the tubero-infundibular region. Very strong evidence indicates that dopamine is a neurotransmitter in some neurons of the central nervous system.

1.1.4. Conversion of Dopamine to Noradrenaline

In 1955, it was shown that the adrenal medulla could convert dopamine-C^{14} to noradrenaline (NA). The enzyme responsible was dopamine-β-hydroxylase in the presence of ascorbic acid and oxygen, and it could be purified from chromaffin granules of the adrenal medulla. At one time, it was thought that dopamine-β-hydroxylase was the rate-limiting step in the biosynthesis of noradrenaline, but Levin and Kaufman (1961) showed that there was enough enzyme present in the adrenal medulla to form all the NA and adrenaline (A) in less than 15 sec. Not much is known about dopamine-β-hydroxylase except its distribution. Udenfriend and Creveling (1959) found it in the adrenal medulla, the hypothalamus, and certain other brain areas. It is localized within noradrenaline storage granules at neuronal synaptic vesicles (Potter and Axelrod, 1963) and in adrenal chromaffin granules (Kirshner, 1959). Almost all the dopamine taken up by neuronal NA vesicles is transformed within the vesicle to NA. The enzyme is also nonspecific in that it can hydroxylate tyramine, 3-phenylethylamine, and N-methyldopamine (epinine) on the β-carbon.

1.1.5. Conversion of Noradrenaline to Adrenaline

The N-methylation of NA occurs under the influence of S-adenosyl methionine methyltransferase with the cofactor S-adenosyl methionine, which forms from methionine and adenosine triphosphate (ATP) in the presence of magnesium ions.

This enzyme is highly localized in the soluble supernatant fraction of the adrenal medulla as well as in brain. The enzyme can also N-methylate a number of normally occurring derivatives such as normetadrenaline, octopamine, β-phenylethanolamine, and foreign phenylethanolamine derivatives. In the adrenal medulla, the metabolism of catecholamines takes place through a series of intracellular displacements of substrates as follows: L-dopa passes through the cytoplasm, where it is decarboxylated; the dopamine that has formed passes through chromaffin granules, where it is transformed into NA; the NA returns to the cytoplasm and is N-methylated to A; finally, A returns to the chromaffin granules, which become its place of storage until liberation.

1.1.6. Biosynthesis of Catecholamines in the Brain

The biosynthesis of catecholamines in the brain requires the presence of tyrosine, a naturally occurring amino acid. Tyrosine is present in the blood in concentrations of 8×10^{-5} M (Ikeda et al., 1966) and is transported to the brain, where it is converted to L-dopa by the enzyme tyrosine hydroxylase. Although a lesser pathway has been shown in the production of noradrenaline through the amino acid tyramine (Creveling, 1962), the urinary yield of noradrenaline and normetadrenaline is usually small. This second alternative pathway of tyramine to octopamine or dopamine (Fig. 3) is thought to occur mainly in the liver rather than in nervous tissue. The presence of tyrosine hydroxylase in the brain, shown by Nagatsu *et al.* (1964), makes this enzyme the rate-limiting step in catecholamine synthesis, and it selectively oxidizes L-tyrosine to L-dopa in the presence of oxygen. Drugs such as α-methyltyrosine can effectively inhibit the enzyme and thus interfere with the normal synthesis of catecholamines in the brain. Patients given α-methyltyrosine show a characteristic clinical picture of reduced motor activity and lethargy, but no muscular atony or Parkinsonian-like tremor as can be expected from drugs which reduce the levels of dopamine in the brain. After the hydroxylation of tyrosine and the production of dopa, the latter amino acid is decarboxylated to form dopamine by the nonspecific enzyme dopa decarboxylase. Dopa decarboxylase requires pyridoxal phosphate as a cofactor. A series of decarboxylase inhibitors is known, some of which inhibit the enzyme peripherally and others endogenously. The peripheral decarboxylase inhibitors can potentiate the penetration of administered L-dopa into the brain parenchyma, while endogenous decarboxylase inhibitors such as α-methyldopa can reduce the concentration of catecholamine stores, not exactly by inhibiting the decarboxylase enzyme, as was previously believed, but rather by decreasing noradrenaline formation in the brain by decarboxylation to α-methyldopamine, which sub-

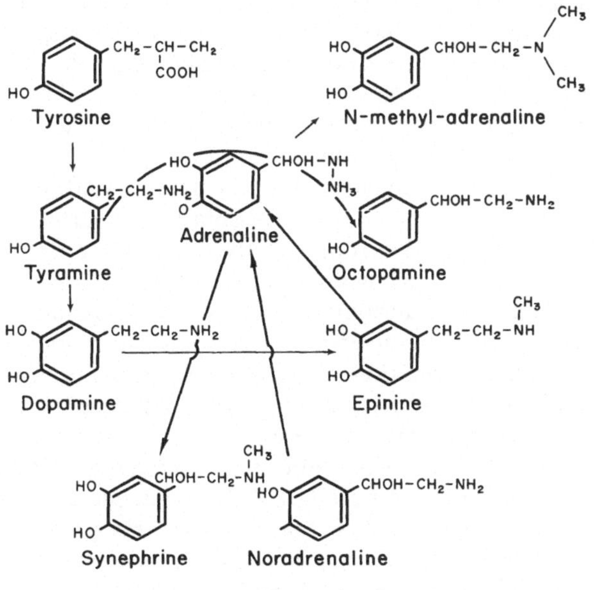

Fig. 3. Alternate metabolic pathway to adrenaline. Metabolites such as octopamine, epinine, synephrine, and *N*-methyladrenaline are found as naturally occurring in nature.

sequently is converted to α-methylnoradrenaline, which depletes endogenous noradrenaline from its intracellular stores in sympathetic neurons (Carlsson and Lindquist, 1962). The conversion of dopamine to noradrenaline in the brain is catalyzed by dopamine-β-hydroxylase, a copper-containing enzyme found within the vesicles of nerve endings. This enzyme requires the presence of oxygen and ascorbic acid as cofactors and can be inhibited by disulfiram *in vitro* and *in vivo*, thus lowering central levels of noradrenaline. In theory, therefore, ascorbic acid deficiency could clinically reduce the enzyme activity, but attempts to demonstrate this *in vivo* have failed. On the other hand, chelating agents which bind copper are effective inhibitors of this enzyme. Goldstein and Nakajima (1967) have shown that the action of disulfiram on dopamine-β-hydroxylase comes about through the formation of diethyl dithiocarbamate, a reduced product of disulfiram by ascorbic acid which can effectively chelate copper. It is of interest to note that Missala *et al.* (1967) has shown that copper-deficient animals have a reduced conversion of tyrosine to noradrenaline. Besides the decrease in noradrenaline caused by the inhibition of dopamine-β-hydroxylase, there is a concomitant increase of dopamine in the brain. The final step in the conversion of noradrenaline to adrenaline is carried out through the enzyme phenylethanolamine-*N*-methyltransferase. It is possible that noradrenaline cannot leave the brain substance without prior deamination by monoamine oxidase, indicating that a blood–brain barrier mechanism works in both directions. Experiments by Glowinski (1967) indicate that radioactive endogenous noradrenaline can be synthesized in the brain and particularly in nerve endings, as shown by the injection of radioactive tyrosine and dopa as precursors (Glowinski and Iversen, 1966*b*). These same experiments proposed that dopamine injected into the lateral ventricles accumulates in both noradrenaline- and dopamine-containing neurons. However, the noradrenaline-containing neurons rapidly convert the dopamine to noradrenaline, while regions rich in dopamine content, such as the neostriatum, do not, suggesting that the terminals there lack the dopamine-β-hydroxylase enzyme necessary for the conversion of dopamine to noradrenaline. There is, however, a small formation of noradrenaline in the neostriatum, and information concerning the metabolism of noradrenaline and dopamine in that region indicates the presence of some noradrenaline-containing nerve endings in addition to the numerous dopamine-containing synaptic terminals. *O*-methylation of noradrenaline can occur in the brain as well as in extracerebral organs such as the liver, and this process appears to be extraneuronal. This would suggest that catechol-*O*-methyltransferase (COMT), which catalyzes the *O*-methylation sequence, probably acts on the amines after they have been liberated from the nerve cells. It should be pointed out, however, that the exact location of the action of COMT is not known but that COMT is believed to be located outside the neurons, where it can *O*-methylate noradrenaline either in this tissue itself or in the liver when noradrenaline is carried there by the circulation.

In subsequent experiments, Glowinski *et al.* (1966*b*) have found that amphetamine and desmethylimipramine inhibit the uptake of ventricularly injected noradrenaline and that the inhibition is stronger in the regions with a high turnover of this amine. Noradrenaline transport may be different from dopamine transport since these two drugs block the uptake of noradrenaline but not of dopamine. There is strong evidence to suggest that the synthesis of catecholamines (i.e., dopa to dopamine, to noradrenaline) occurs in the nerve terminals of the mammalian brain. In support of this hypothesis, it has been reported that after intraventricular injection of dopa-H^3 there is found some hours later radioactive dopamine and noradrenaline in the particles isolated from brain homogenates. These particles belong to the synaptosomes or pinched-off nerve endings. The subcellular metabolism and localization of the monoamines will be discussed at more length in Chapter 3.

The storage and release of catecholamines at nerve endings have been the subject of numerous recent studies. The evidence indicates that storage of DA and NA is dependent on two types of mechanisms: (1) uptake at the neuronal level and (2) binding. The process of catecholamine uptake simply involves the entry of either DA or NA into the cell. *Binding* refers to the amine's ability to enter endogenous vesicles located in nerve terminals which protect them from enzymatic destruction. It is possible that aminergic synthesis can occur anyplace intraneuronally, although the fabrication of storage granules may take place only in the neuronal cell body. From there, axoplasmic flow carries the granules to presynaptic terminals, where they accumulate and can be seen by electron microscopy as dense core vesicles of variable diameter. The release of NA therefore is achieved only from the terminals. Monoamine oxidase (MAO) is also localized within the terminal in the mitochondrial fraction and can regulate the concentration of uptake and release of NA by deaminating the unbound portion of NA. It is obvious that if MAO is inhibited by drugs, the NA concentration in tissue will increase. Similarly, if the uptake of NA is blocked or if the storage granule is stimulated to release supernormal quantities of NA, then the tissue NA will be decreased, with a concomitant increase in the order of its deaminated or methylated products (3,4-dihydroxymandelic acid and normetadrenaline, respectively).

Normally, a steady state exists in the uptake, storage, release, and regulation of NA. It follows, then, that if tyrosine hydroxylase is the rate-limiting step in the synthesis of catecholamines, any mobilization of NA from its active pool would also reflect some changes in the levels of tyrosine hydroxylase. This, however, does not occur; at least after cold stress or during exercise, when the NA synthesis is increased, no changes in tyrosine hydroxylase activity are observed. It has been suggested that NA molecules in the nerve terminal may regulate the activity of tyrosine hydroxylase, perhaps by competing with the pteridine cofactor. McGeer and McGeer (1967) have shown in *in vitro* studies that various catechols, DA, and NA can in fact inhibit tyrosine hydroxylase. Besides competing for the pteridine cofactor, NA mediation of tyrosine hydroxylase could conceivably occur through its interaction with a separate, regulatory binding site on the enzyme. The question of how NA release is terminated or shut off is also open to speculation. Possibly, negative feedback mechanisms of end products or active uptake of NA regulate the amines' rate of synthesis. The release and fate of DA and NA in brain are similarly complex.

Catecholamine turnover is the process involving the disappearance of amines from the nerve cell. This phenomenon is evident in the release of NA from terminals. Acetylcholine (ACh) has been suggested as the necessary substance for the release of NA in postganglionic fibers (Burn and Rand hypothesis). Briefly, a modification of this hypothesis (Burn, 1966) argues that both ACh and NA are localized in the same terminal and that the former facilitates the release of NA, which then acts on the effector cell. However, the most deficient part of this argument is the fact that after iontophoretic administration of ACh on noradrenergic neurons, no NA is released. Moreover, release of ACh or NA from cells may be achieved selectively without involving the release of both substances. Once released from its storage sites, NA can stimulate central receptors, with part of it being inactivated either intraneuronally by MAO or extraneuronally by both MAO (organ bound) and COMT. Experiments with NA-H^3 show that NA can be taken up by the tissue within 2 min after its injection. This rapid uptake may be a partially protective insurance for NA, since circulating COMT can destroy it or, if it is taken up by an organ (e.g., liver), it can be degraded by either MAO or COMT.

The uptake of exogenous amine by a noradrenergic cell depends, among other things, on the amount of NA available. Such uptake of DA, NA, and serotonin in brain has been well studied using the histochemical fluorescence technique. With this technique, lipid-soluble monoamine analogues capable of crossing the blood–brain barrier have been studied in relation to their uptake at cell membranes (Carlsson, 1966). The displacement of intraneuronal amines provoked by these analogues is another tool used in providing information on their uptake, binding, and release mechanisms.

The schematic representation of an noradrenergic nerve terminal in Fig. 4 shows the steps involved in the uptake, synthesis, binding, and release of NA from its storage granules. Electron microscopic evidence (De Robertis, 1966; Whittaker, 1966) indicates that in adrenal medulla, in various splenic nerves, and in the central nervous system, certain osmophilic granules exist. The diameter of these granules varies from 0.05 to 0.2 μ,

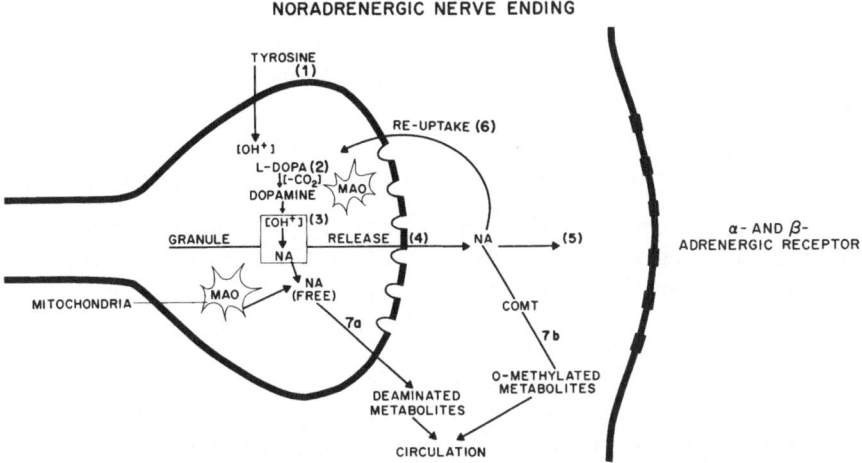

Fig. 4. (1) Tyrosine is hydroxylated in cytoplasm to L-dopa; (2) L-dopa is decarboxylated to dopamine (cytoplasm); (3) dopamine enters granule and is hydroxylated to noradrenaline; (4) noradrenaline is released; (5) adrenergic receptor is stimulated; and (6) active reuptake mechanism captures some of the extracytoplasmic noradrenaline. (7a–7b) Deamination by MAO or O-methylation by COMT of the free NA may occur.

and they probably contain high concentrations of catecholamines and adenosine triphosphate (ATP) in a ratio of 4:1.

The synthesis of tyrosine to L-dopa occurs within the cytoplasm, where further decarboxylation results in the formation of DA. The final step in the synthesis takes places as DA enters the granules, where dopamine-β-hydroxylase acts on the amine to produce NA. The second major aspect of NA increase at the nerve terminal is the active reuptake mechanism after its release from its storage sites. This reuptake phenomenon is postulated to terminate the effects of adrenergic impulses following NA release. The exact mechanism responsible for the sequence of steps leading to this selective release of NA from adrenergic fibers is not presently known.

For further information about the most recent work on the aspects of storage and release of the catecholamines, the reader is referred to the recent *Bayer Symposium* (Schumann and Kroneberg, 1970) dealing with this subject.

1.2. BIOCHEMISTRY OF 5-HYDROXYTRYPTAMINE (SEROTONIN)

Several pathways are known which lead to 5-hydroxytryptamine (5-HT), or serotonin, beginning with certain substrates present in some enzymatic systems. The minor collateral metabolic steps leading to serotonin are still not completely understood and will require further study before they can be fully assessed. Instead, the major metabolic pathway will be discussed.

1.2.1. Conversion of Tryptophan to 5-Hydroxytryptophan

The enzyme responsible for the conversion of tryptophan to 5-hydroxytryptophan (5-HTP) is tryptophan-5-hydroxylase, a substance found in mammalian tissue and the rate-limiting enzyme in the formation of serotonin (Figs. 5 and 6).

Administration of *dl*-tryptophan-2-C^{14} has shown its conversion to labeled 5-HTP recovered from liver homogenates in the rat and guinea pig. The significance of these results has been questioned, however, because only a small fraction of the original radioactivity was found in the 5-HTP.

Attempts to purify tryptophan-5-hydroxylase had been unsuccessful until Cooper and Mercer (1962) reported finding the enzyme in a particulate fraction of intestinal mucosa and kidney. Clinically, the hydroxylation of tryptophan to 5-HTP has been shown by Perry *et al.* (1964) and Baldridge *et al.* (1959) to be defective in patients with phenylketonuria. After an oral load of tryptophan in these patients, only slight increases in serotonin

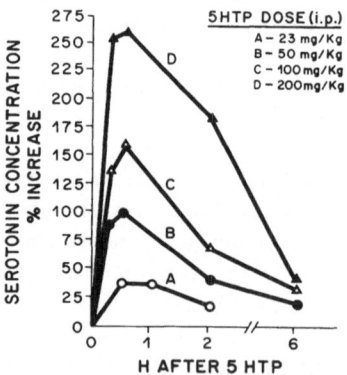

Fig. 5. Effect of 5-HTP on rat brain serotonin concentration. Each point is an average of three separate experiments and is compared to average value from a pool of six normal brains. Normal brain concentration (μg/g) from nine experiments (average $\pm$ SEM) was 0.534 $\pm$ 0.012. From Green and Sawyer (1964).

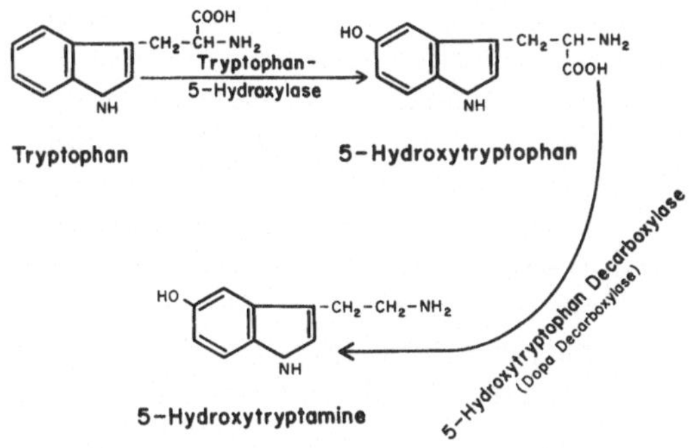

Fig. 6. Synthesis of 5-HT from tryptophan.

and 5-hydroxyindoleacetic acid were found, suggesting a pathochemical element in the hydroxylation of tryptophan similar to that of phenylalanine conversion blocking to tyrosine characteristic of phenylketonuria. It is possible that the same enzyme may be involved in the liver hydroxylase activity of both tryptophan and phenylalanine. Lovenberg *et al.* (1968) has shown evidence that phenylalanine in brain may inhibit tryptophan hydroxylase activity, but the possibility of other mechanisms influencing this effect has not yet been ruled out.

Localization, purification, and demonstration of the properties of tryptophan-β-hydroxylase in the brain by Grahame-Smith (1964) and later its demonstration by assay procedure by Green and Sawyer (1966) substantially clarified the role of this enzyme in the biosynthesis of serotonin. Tryptophan hydroxylase is the rate-limiting step in the formation of serotonin. The enzyme can be inhibited *in vitro* and *in vivo* by p-chlorophenylalanine, thus reducing the brain level of serotonin to about 14% of the original amount without drastically changing the catecholamine content. The first evidence that tryptophan-5-hydroxylase could be located in the brain came after experiments by Gal and Marshall (1964) following direct administration of radioactive tryptophan into the brains of rats and pigeons. This group found that labeled tryptophan given to pigeons intraperitoneally did not result in any labeled serotonin in the brain, suggesting that the precursor was taken up by other tissues and synthesized by a separate system. The logical conclusion was that the brain contains its own biosynthetic mechanism to synthesize from tryptophan by hydroxylation the serotonin precursor 5-HTP and then serotonin. Other indirect evidence appeared to support the theory that serotonin is synthesized in the brain. If, for example, the gastrointestinal tract, which is the major site of serotonin synthesis in the body, was removed, there resulted a decrease of serotonin in the serum, lungs, and spleen, but there was little effect on the brain serotonin. There was also unpublished evidence obtained by Garattini of an increase of hydroxyindoles in brain after intracerebral administration of tryptophan. The probable site of tryptophan hydroxylase in the brain has been shown by Grahame-Smith (1964), Nakamura *et al.* (1965), and Gal (1964) to be exclusively in the particulate fraction of brain homogenates, probably in the mitochondria. Ultracentrifugation studies have confirmed that the hydroxylase enzyme is found to a great extent in nerve endings and also in the denser mitochondrial fraction. It has been detected only in very small amounts in the microsomal, nuclear, or soluble fractions. This is unlike 5-HTP decarboxylase, which is found in the soluble, microsomal, and crude mitochondrial fractions.

After the hydroxylation of tryptophan to 5-HTP, the pathway proceeds to formation of serotonin by a catalytic process from the same nonspecific decarboxylase which converts dopa to dopamine. This enzyme has been called dopa decarboxylase, or aromatic L-amino acid decarboxylase. A pteridine cofactor is also necessary in the hydroxylation process from tryptophan. The presence of this cofactor is also necessary in the tyrosine hydroxylase synthesis. This similarity led Udenfriend *et al.* (1965) to show that the catechol compounds can exhibit a feedback mechanism control on tyrosine hydroxylase due to competition between the catechol and the reduced pteridine cofactor for a binding site on the enzyme. Pineal tryptophan hydroxylase is also strongly inhibited by catechols. The anatomical localization of tryptophan hydroxylase and that of L-aromatic acid decarboxylase in the brain of the guinea pig seem to closely parallel each other. We will see in later sections how the use of some drugs such as *p*-chlorophenylalanine can affect the concentration of serotonin in the brain, and we shall discuss the role of this indoleamine in the central nervous system.

1.2.2. Conversion of 5-HTP to 5-HT

Work by numerous investigators has shown that 5-HTP and dopa are decarboxylated by the same nonspecific enzyme, 5-HTP decarboxylase (or aromatic L-amino acid decarboxylase). The enzyme has been found in brain of rats, guinea pigs, pigeons, mice, rabbits, hogs, dogs, and humans. Administration of 5-HTP results in some increase in brain serotonin, a fact which suggests that the transport mechanism for the enzyme is similar to that for tryptophan, which is also taken up by brain through active transport. Pletscher and Gey (1962*a*) have found decarboxylase activity in brain to continue for several hours after the death of the animal. The decarboxylase has been reported by Giarman (1956) to be present in the nonparticulate fraction of the cell.

The distribution of 5-HTP decarboxylase has been reported highest in the nucleus caudatus and hypothalamus, while the lowest values are found in the cerebellum and cerebral cortex. In rat brain slices, the transport of 5-HTP has been shown to increase through contact with oxygen and glucose but to be reduced by dinitrophenol if pH is optimum. Since 5-HTP decarboxylase is well distributed throughout the organism, the administration of 5-HTP will result in the formation of serotonin at sites where this indoleamine is normally absent.

Serotonin level in the brain appears to be in direct relationship to 5-HTP decarboxylase concentration, and, as Bogdanski *et al.* (1956) have pointed out, the amine–enzyme levels are closely matched from one region of the brain to another. The only exceptions to this concentration relationship are found in the pyriform cortex, caudate nucleus, and amygdala, where the amine and the enzyme decarboxylase follow masses of gray matter along the polysynaptic descending pathways described by Gloor (1955*a*, *b*) which originate in the amygdala and finally converge on the hypothalamus.

1.3. INACTIVATION OF MONOAMINES

The chemical degradation process of catecholamines is dominated by two enzymatic systems which are fundamental: catechol-*O*-methyltransferase (COMT) and monoamine oxidase (MAO). Recent evidence appears to indicate that *O*-methylation of noradrenaline occurs not only in the liver but also in the brain, where the amines are catalyzed by COMT extraneuronally, that is, after amine liberation from the neurons.

1.3.1. Catechol-*O*-Methyltransferase

Catechol-*O*-methyltransferase is an enzyme found in all tissues containing catecholamines. Its cofactor is *S*-adenosyl methionine. It may be inhibited by pyrogallol, which acts in competitive inhibition with the natural substrate. This last is made up of various catechols and catecholamines which under the action of the enzyme are methylated by a hydroxyphenol in the *meta* position. The distribution of COMT is widespread. It is found in tissues, glands, blood vessels, sympathetic and parasympathetic nerves and ganglia, and all areas of the brain. It has been found also in fish and amphibian tissues. In brain, its highest activity is in the area postrema and its lowest in the cerebellar cortex. Catechol-*O*-methyltransferase is confined mainly to the soluble supernatant fraction and, in smaller amounts, the microsomal fraction of the liver.

1.3.2. Monoamine Oxidase

Monoamine oxidase has been defined as an enzyme which replaces the amino group of a substance with a carbonyl group, does not attack α-methylamines, and is not inhibited by semicarbazides. The exact function of MAO has not yet been resolved. It is thought by some to be concerned with the deamination of O-methylated derivatives of amines and by others to be a substance which participates in the control of intracellular stores of catecholamines. A dual function has also been suggested. The enzyme is localized in the mitochondria of cells in the nerve endings and is well distributed in the organism, particularly in the liver, kidneys, and intestines. It is also found in high concentrations in sympathetic nerves. Recent studies made on organs rich in sympathetic nerve endings, such as the pineal gland, show that upon sympathetic denervation of the organ, MAO is markedly reduced.

Some monoamines appear to be better substrates for MAO than others. Kopin (1964) has shown that although noradrenaline and adrenaline can serve as MAO substrates, it would seem that MAO does not play a major role in the inactivation of these two circulating catecholamines. Dopamine, however, can be O-methylated and becomes a far better substrate than the β-hydroxylated catecholamines. Further work has revealed that catecholamines are first enzymatically attacked by COMT and later acted upon by MAO. On the other hand, MAO seems to play a more active role with cerebral tissue amines than COMT, a detail which suggests its importance in brain amine metabolism. MAO can catalyze the oxidative deamination of tyramine, tryptamine, and serotonin as well as the catecholamines.

A number of hydrazines are capable of inhibiting MAO, among them iproniazid, nialamide, phenelzine, phemiprazine, and phenoxypropazine. *In vitro,* these inhibitors may compete with MAO along with other substrates as long as inhibition is not complete. If inhibition is complete, then the reaction is irreversible and noncompetitive. The inhibition of MAO in the intact animal potentiates the physiological effects of tyramine but not those of noradrenaline or adrenaline. This effect may be due to the rate of disappearance of tyramine when the circulation is slowed down and thus becomes more available to liberate catecholamines from the various tissues.

MAO can also transform monoamines into aldehydes, which can be further oxidized to acids or reduced to alcohols, as shown in Fig. 7. The reduction process predominates in the rat, while the oxidation process is more common in man.

Serotonin is liable to enzymatic action by MAO when it is formed or released from granules. At present, MAO seems to be the only enzyme capable of metabolizing 5-HT in brain. Monoamine oxidase activity is well

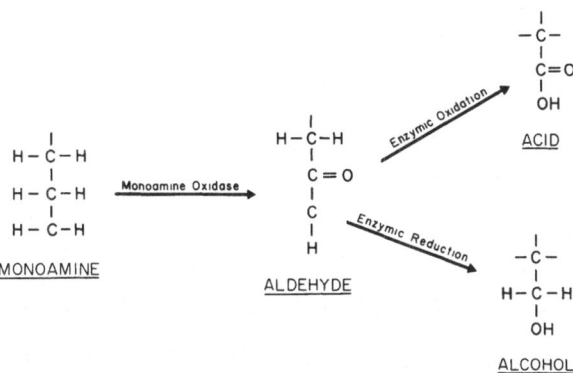

Fig. 7

distributed in brain and ranges from a high threefold activity in the hypothalamus to a low in the corpus callosum. Bogdanski *et al.* (1956) have also suggested the possibility of the existence of more than one MAO, the other being more specific for serotonin. This phenomenon would be similar to the relative specific and nonspecific enzymes involved in the metabolism of acetylcholine. The distribution of serotonin in various cytoplasmic fractions of rat brain has been found to be exclusively integrated in the mitochondrial fraction of cells. In the brain, MAO is present in the neuronal and synaptic mitochondria and has been found within the brain capillaries, where it deaminates the monoamines before they can pass through the blood–brain barrier. The COMT enzyme apparently has a different topography than MAO does, although its exact location within the synaptic region has not yet been determined.

The field of MAO inhibitors has opened the door to modern neuropharmacological therapy, with more than 500 inhibitors so far described. Nevertheless, the precise action and chemical configuration of MAO will remain simply theoretical until the enzyme is sufficiently purified to permit definite evaluation and understanding of its functional significance in the living organism.

Chapter 2

Methods of Studying Cerebral Monoamines and Their Enzymes

2.1. EVALUATION OF METABOLITES EXCRETED IN URINE

Most of the major metabolic products of catecholamines and indole-amines are now well known. They include 4-hydroxy-3-methoxymandelic acid (HMMA), homovanillic acid (HVA), normetadrenaline, metadren-aline, 3-methoxytyramine, and 4-methoxyphenylglycol (HMPG). Other minor metabolic products have also been shown: 3,4-dihydroxymandelic acid (DHMA), 3,4-dihydroxyphenylglycol (DHPG), *N*-methyladrenaline, and *N*-metamethyladrenaline (Fig. 8).

The major catecholamine metabolites were first recognized by Axelrod (1957) and Armstrong *et al.* (1957) using a paper chromatography technique. This technique is useful for detecting and estimating HVA, metadrenaline, 3,4-dihydroxyphenylacetic acid, normetadrenaline, DHMA, and HMMA. Most of the metabolites are estimated from urine, but some can only be assayed by tissue analysis. In urine determinations, the sample is either assayed directly or after preliminary acid hydrolysis if the metabolite has been excreted as a conjugate. Excretion of hydroxyindoles is also determined clinically, most often from 5-hydroxyindoleacetic acid (5-HIAA). Testing for HIAA in urine is one way of calculating the amount of serotonin metabolized in the body. The 5-HIAA is eliminated by glomerular filtration and tubular secretion. The elimination of this metabolite has been found by Johnsen *et al.* (1958) to be subject to diurnal variation, but this observation is not supported in the study by Langemann and Goerre (1957), who found no change in 5-HIAA excretion during night or day. The removal of the large intestine in rats, dogs, and humans has been shown to decrease the urinary excretion of 5-HIAA, and if the gastrointestinal tract is completely removed the level of serotonin after 3 days is also lowered by 90% in serum, spleen,

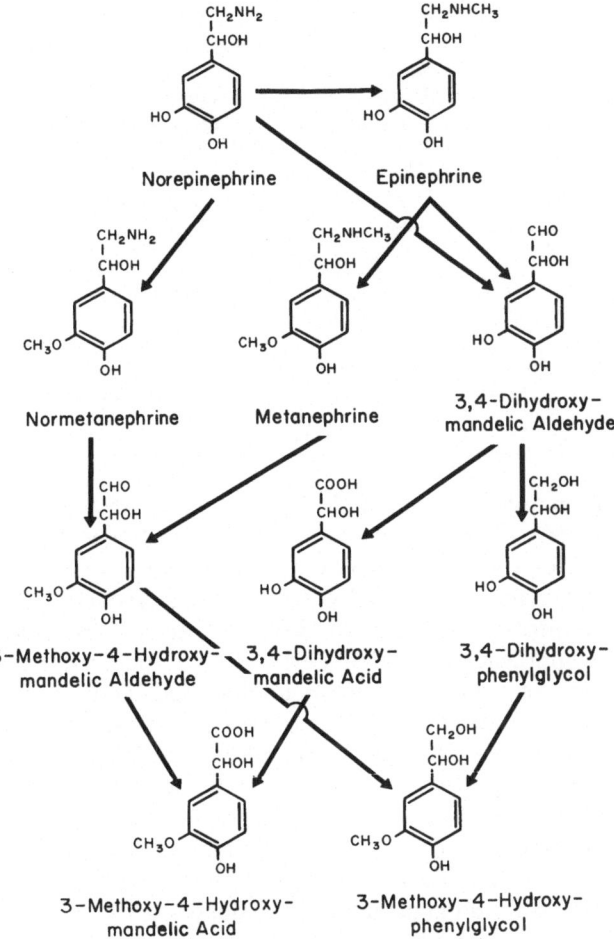

Fig. 8. Catecholamine metabolites.

and lung but not in skin or brain. A deficiency of the serotonin precursor tryptophan also causes a 50% fall of HIAA excretion in urine after 3 days. Species difference in hydroxyindole excretion has also been observed. In the rabbit, horse, and guinea pig, 5-HIAA is excreted in relatively small amounts, while in the dog, rat, and human, the excretion rate is more variable but the content somewhat higher. Table 1 shows the excretion values of some of the species studied.

Studies in rats show that the excretion value of 5-HIAA in urine depends on whether the animal has received 5-HT or 5-HIAA. If 5-HT is administered, the recovery percentage of 5-HIAA (Table 1) ranges from 38 to 33%

Table 1

Species	Normal value of 5–HIAA excretion (μg/ml)	Percent of administered 5–HT excreted as 5–HIAA
Dog	2.5–4	25
Guinea pig	0.3	1
Hog	1.5	—
Horse	0.3	—
Man	2–3.8	—
Kid	1.25	20
Lamb	2.25	—
Ox	0.3	—
Rabbit	0.3	1
Rat	1.2–1.5	38–33
Toad	0.27	—

of the serotonin injected, but if 5-HIAA is given instead, the recovery percentage is 76% of the indole acid administered. The 5-HIAA is considered the major metabolic product of serotonin catabolism, but also recovered from urine are N-methyl-5-HT, N-acetyl-5-HT, O-sulfate, O-glucuronide conjugates, 5-O-sulfate indoleacetic acid, 5-hydroxytryptophol, and 5-hydroxyindoleaceturic acid. Certain other indole metabolites have also been found in urine but remain unidentified. O-aminobenzoate, an inhibitor of glucuronide formation, increases the urinary excretion level of 5-HIAA when given to rats after 5-HT, but not after 5-HIAA administration. Donaldson et al. (1959) concluded from the above observation that serotonin, but not 5-HIAA, is conjugated. Other studies indicate that the fall of 5-HIAA excretion after treatment with MAO inhibitors is related in some way to an increase of 5-HT-O-sulfate. That being the case, the formation of the 5-HIAA found in urine would be dependent on tissues metabolizing 5-HT through the action of MAO.

During some analyses of human urine, von Euler and Lishajko (1959) found that acid hydrolysis of the urine caused a large increase in the amount of catecholamine, presumably by the splitting of a conjugate.

The study of monoamine metabolites in urine has brought about in recent years a great many different techniques that try by qualitative and quantitative measurements to correlate excretion factors with normal and pathological states. Urine studies are usually preferred in lieu of blood or tissue samples as the amine concentrations in the latter two are very low and detection or measurement is often difficult. Amines in blood and tissue are also rapidly destroyed by MAO and other enzyme systems, but small amounts are nevertheless excreted and may be shown in urine. The value of measuring the increased catecholamine excretion in the diagnosis of pheochromocytoma and serotonin excretion in argentaffinoma is clinically well established. Smaller variations in catecholamine excretion have also been reported in cases of psychic and physical stress.

2.2. LABELING AND AUTORADIOGRAPHIC STUDY OF MONOAMINES

Monoamines may be studied both *in vivo* and *in vitro* by such techniques as isotopic labeling and autoradiography. The first technique involves using a labeled molecule, generally C^{14} or H^3, which is administered to man or animal and its action (fixation, transformation, elimination) then studied in various tissues, blood, or urine. The molecule and/or its metabolites are recovered by extraction with organic solvents and separated by column chromatography, thin-layer chromatography, etc. The radioactivity of the different isolated fractions can then be measured and analyzed accordingly.

Since Blaschko's (1939) discovery of the major pathway to adrenaline from tyrosine in the adrenal medulla, investigation of catecholamine distribution has also revealed the presence of amines in sympathetic nerves and central nervous system. From experiments following the intravenous administration of noradrenaline-H^3 and adrenaline-H^3, Axelrod (1962) concluded that both amines are rapidly but unequally taken up by heart, spleen, and glandular tissues, with skeletal muscle tissues showing the least uptake. Also, the high catecholamine uptake by the heart indicates that catecholamines are discharged by the adrenal gland and sympathetic nerve endings into the bloodstream, where they serve to supply the heart. It was observed that 2 hr after the administration of H^3-labeled catecholamines, the concentrations of these amines in heart, spleen, and adrenal gland were the same as after 2 min. This indicates that NA and A are retained in a physiologically active form until they are released by the tissues. This retention or binding to tissues may protect the amines from enzymatic attack and could function as an important mechanism for storing and inactivating these hormones.

Regional studies of H^3-labeled catecholamines indicate that very small amounts can be taken up by the hypothalamus but that the rate and quantity of uptake are much less than in extracerebral tissues. Axelrod (1962) has proposed that MAO, being higher in rat brain than COMT, is mainly concerned with the inactivation of noradrenaline in the central nervous system. Adrenaline, on the other hand, is more dependent on enzymatic O-methylation for storage and discharge. Noradrenaline-C^{14} administered directly into the lateral ventricle of the cat is metabolized in brain, with products such as normetadrenaline, HMMA, and 4-hydroxy-3-methoxy-

this amine is localized in nerve terminals of brain, with smaller amounts in axons and cell bodies. After intraventricular injection, the radioactive NA accumulates in the central adrenergic neurons, mixing with the endogenous store of catecholamines. The subcellular distribution of tritium-labeled dopa, dopamine, and NA in the rat brain in particulate and supernatant fractions (P/S ratio) has been studied by injecting these labeled catechol-amines into the lateral ventricle. After injection of NA-H^3, a close correla-tion in distribution was found between the radioactive NA and its unlabeled endogenous counterpart in the following regions: In the cerebellum, endog-enous and exogenous NA had the lowest P/S ratio, while it was highest in the hypothalamus. In the cortex, hypothalamus, and medulla, the P/S ratio of NA-H^3 4 hr after injection was lower than the P/S ratio of endog-enous NA. In the striatum, endogenous dopamine was predominantly recovered in the S fraction, and exogenous dopamine-H^3 and NA-H^3 were found mainly in the P fraction. After administration, NA-H^3 was detected in the synaptosomal layer, striatum, and hypothalamus. Dopa-H^3 and synthesized dopamine and NA were also recovered in the synaptosomal layer after density gradient centrifugation of striatal or hypothalamic homogenates.

Tracer techniques for the study of serotonin in brain employ precursors of this amine since intravenously injected tritiated 5-HT itself does not penetrate the blood–brain barrier in appreciable amounts. For instance, tritiated 5-HT may be introduced into brain by parenteral injection of its precursor, tritiated 5-HTP. In *in vitro* studies, nerve ending particles have been incubated with C^{14}-labeled 5-HTP in a medium which contains brain supernatant fraction and in which the C^{14}-labeled 5-HTP is de-carboxylated to C^{14}-labeled 5-HT. Less 5-HT is taken up by the nerve ending particles when C^{14}-labeled 5-HTP is decarboxylated to C^{14}-labeled 5-HT and less 5-HT is taken up by the nerve ending particles when C^{14}-labeled 5-HTP is replaced by C^{14}-labeled 5-HT, facts which suggest that the addi-tion of the supernatant fraction decreases the uptake of radioactivity by the nerve ending particles (regardless of whether they are incubated with labeled 5-HT or 5-HTP). The nature of this interference by the supernatant fraction is not known, but there is a possibility that acidic substances in the fraction bind the amine before the particles do. Another possibility may be that the supernatant fraction or the 5-HTP participates in a regulatory mechanism that controls *in vitro* the extent of serotonin binding to the particles.

Radioactive tryptophan has also been used by Gal *et al.* (1964) to study the effect of this serotonin precursor in pigeon and rat brain. They found that 20–30 min after injection of L-tryptophan-C^{14} into the subarachnoid space of pigeons, C^{14}-labeled 5-HT could be recovered from brain tissue. Intra-peritoneal injection of the same labeled tryptophan did not yield radioactive 5-HT in brain even when the dosage of tryptophan was increased 20 times.

In another study complementing this one, intracerebral injection of L-tryptophan-C^{14} led to the synthesis of C^{14}-labeled 5-HT in the rat brain. This evidence indicates that the brain is capable of synthesizing serotonin through a process of hydroxylating tryptophan. The enzyme responsible for this hydroxylation in the intestines is tryptophan-5-hydroxylase, recently detected in brain tissue by Gal *et al.* (1964). It seems unlikely, however, that brain serotonin synthesis could be entirely dependent on this mechanism; rather, the 5-HT found cerebrally would rely on the active transport of 5-HTP and tryptophan released into the general system and also on the cerebral hydroxylation of available tryptophan. Drugs also influence the binding and storage capacity of brain 5-HT; this will be treated more at length in Chapter 3.

2.3. ULTRASTRUCTURES: ELECTRON MICROSCOPE STUDIES

Discussion of electron microscope studies of ultrastructures moves this presentation into the realm of macromolecular structure and function, which together with physical and chemical techniques constitute the field of molecular biology.

Studies of the nervous system by the application of pharmacological agents and labeled ions were formerly correlated with structural anatomy as revealed by the light microscope, but although the light microscope is an important tool in many ways, it lacks the magnification and resolution to reveal subcellular components. Not until after 1953, with the introduction of the electron microscope, were these components visually demonstrated. The most characteristic structural components of nerve endings are the synaptic vesicles, first shown by De Robertis and Bennett (1955) and later identified in mammalian brain homogenates from the mitochondrial fraction of rat brain. The cell fractionation method consists of gradient centrifugation and sucrose density gradient fractionation (density is expressed as the sucrose concentration, in molar units, in which the particles float). Essentially, the technique separates the centrifuged fractions (for example, of nerve endings) to give a gradient separation of the cellular components (Fig. 10). Cell fractionation can also be used to isolate synaptic vesicles and separate different parts of nerve ending membranes; all components can then be processed for electron microscopic examination. Submitochondrial fractions of brain show that varied levels of monoamines are recovered from specific cell structures. As seen in Table 2, nerve endings separated into two subfractions have an uneven catecholamine concentration, nerve ending type (a) having catecholamine activity and being potentially aminergic and nerve ending type (b) lacking these amines and

Table 3. Certain Characteristics of Noradrenaline Storage Granules

Site	Density	Uptake	Size (Å)	Release by Ca^{2+} (bound NA)
Sympathetic nerve	Same as microsomes	Catecholamines and monophenols	400–500	No
Brain synaptosomes[a]	Between microsomes and mitochondria	Catecholamines and monophenols	1300	—
Chromaffin cell	Heavier than mitochondria	Catecholamines and 5–HT	500–4000	Yes

[a] Synaptosomes isolated from brain tissues are also referred to as "pinched-off nerve endings."

permitted the study of nerve endings and synaptic vesicles in order to discover the subcellular localization of 5-HT. About half of the bound 5-HT is concentrated in the crude mitochondrial fraction, while 40% is found in the microsomal layer. With additional fractionation of the mitochondrial layer, 5-HT appears in the fraction of nerve endings, the same system where acetylcholine (ACh) is present. It is possible, then, that 5-HT and ACh may be contained in the same nerve terminals, or simply that they sediment at the same level although they are contained in different cell compartments.

Preliminary studies of 5-hydroxytryptophan decarboxylase show it to be present in bound form in the crude mitochondrial layer and, like 5-HT, localized in the nerve endings. Hyposmotic shock releases this enzyme totally from its storage site, suggesting that it is loosely stored within the membranes of the nerve ending.

Monoamine terminals in certain brain regions have been studied with the electron microscope by Fuxe et al. (1965), who described the presence of "boutons" similar in size to the varicosities of monoamine terminals. These boutons appeared to be filled with small to medium sized (300–450 Å) vesicles and less common but larger (600 Å) agranular vesicles and to form mostly synapses with adjacent dendrites. Larger granular vesicles (800–1000 Å) were found in all regions studied, and these were not affected by heavy doses of reserpine (catecholamine-depleting agent) or nialamide (MAO inhibitor). Fuxe et al. concluded from these data that monoamine storage granules are not the larger granular vesicles but that they appear as small agranular vesicles due to the technique used.

2.4. HISTOCHEMICAL FLUORESCENCE METHOD

Histochemical fluorescence of amines induced by formaldehyde was first noted in 1952, when cells in the adrenal medulla showed a fluorescence later attributed to noradrenaline (Eränkö, 1955). Lagunoff *et al.* (1961) described a technique for the demonstration of histamine in mast cells by treating extracerebral rat tissues with formaldehyde vapor, thus inducing histochemical fluorescence. At the same time and working independently, a Swedish group (Falck, 1962; Falck and Torp, 1962*a*, *b*; Carlsson *et al.*, 1962*a*; Falck *et al.*, 1962; Bertler *et al.*, 1963*b*; Corrodi *et al.*, 1962) began publishing the details and results of a histochemical fluorescence technique for the demonstration of monoamines (i.e., dopamine, NA, A, and 5-HT) in cerebral and extracerebral tissues. This sensitive and ingenious technique, now well understood thanks to many preliminary studies both in model systems and in diverse vertebrate and invertebrate tissues, has become a highly important tool in the field of monoamine research. The specificity of the reaction permits the visualization of cerebral catecholamines *in situ* by condensation of these amines with formaldehyde vapor in the presence of a dried protein film, which after proper treatment, results in very intense fluorescent products. The reaction for catecholamines proceeds as shown in Fig. 11. Formaldehyde closes the ring by condensation, giving the isoquinoline nucleus which undergoes a secondary reaction to become insoluble and appear as a highly fluorescent green product that has an activation wavelength of 390–410 mμ. Serotonin is seen in the same preparation as a fluorescent yellow product emitting a wavelength of 510–520 mμ, with the condensation reaction following essentially the same pattern as that of catecholamines but giving a β-carboline nucleus (Fig. 12).

No fluorescence develops with the monoamine metabolites, that is, 3-*O*-methylated amines, phenolic acids, tertiary amines, and *N*-acetyl-5-methoxytryptamine (melatonin). Related structures such as phenylethylamines (tyrosine, tryptamine) are also nonreactive, while indolethylamines (tryptophan, tryptamine) are seen as weakly fluorescent products. Monoamine precursors such as dopa and 5-HTP give about the same fluorescence as their corresponding amines. Other compounds correspond-

Formaldehyde

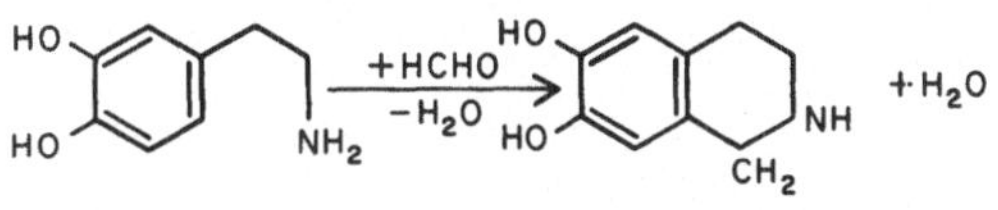

Catecholamine   6,7−Dihydroxy−1,2,3,4 −
Tetrahydro−Isoquinoline

Fig. 11

hold the strips. (3) Specimens are quickly transferred from the cooled propane and immersed in liquid nitrogen; they may remain in the liquid nitrogen until the freeze-drying procedure. (4) Specimens are placed in a freeze-drying apparatus (suitable high-vacuum pump, cooled by dry ice + acetone or cooling coil from compressor-type cooling apparatus) and remain 3–7 days at −30 C throughout the process. (5) On the final day, the temperature in the freeze-dryer is slowly raised from −30 to +35 C in 3 hr and kept at +35 C for another 3 hr, after which the vacuum is broken and the tissues are rapidly transferred to a dry air container containing P_2O_5 or $CaCl_2$, or they may be transferred directly from the dryer to the paraformaldehyde reaction container. Paraformaldehyde powder having a relative humidity of 60–70% (for 65% add to the bottom of a desiccator jar a solution containing concentrated sulfuric acid and water 50:125) is prepared for step 6. (6) Paraformaldehyde powder (5 g/liter volume) is placed in the container holding the tissues (the sulfuric acid solution is removed), and the container is placed in an incubator at +80 C for 1 hr (3 hr for visualizing adrenaline). Formaldehyde vapor is released from paraformaldehyde by heat. (7) After heating, the tissues are embedded in degassed liquid paraffin (48–52 C mp) and sectioned at 10 μ. (8) Cut sections are placed on clean slides and *lightly*

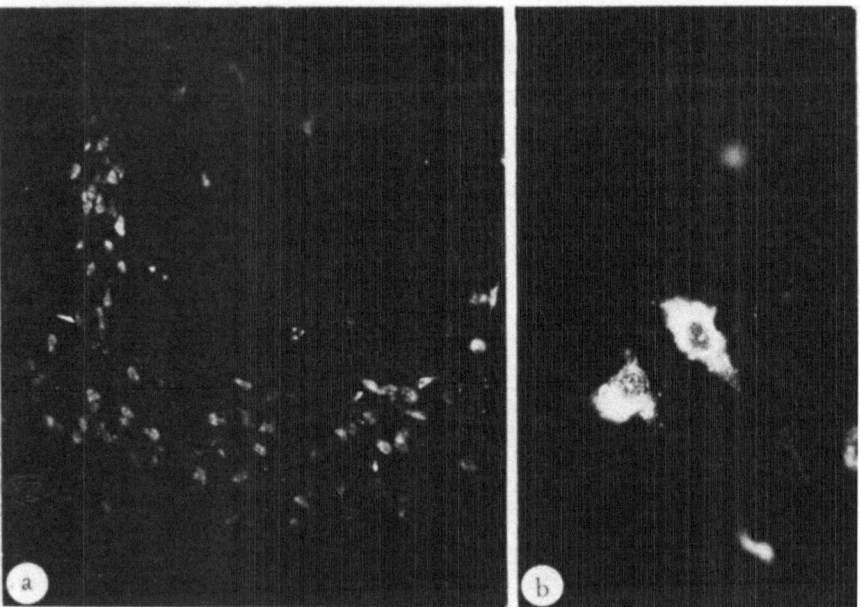

Fig. 15. Substantia nigra. Catecholamine cell bodies (a) surrounding nucleus ruber in the mesencephalic reticular formation. 40×. (b) High magnification (180×) of two cell bodies and their processes from the substantia nigra. Nucleus is nonfluorescent. Rat treated with nialamide (150 mg/kg). (Reduced for reproduction 15%.)

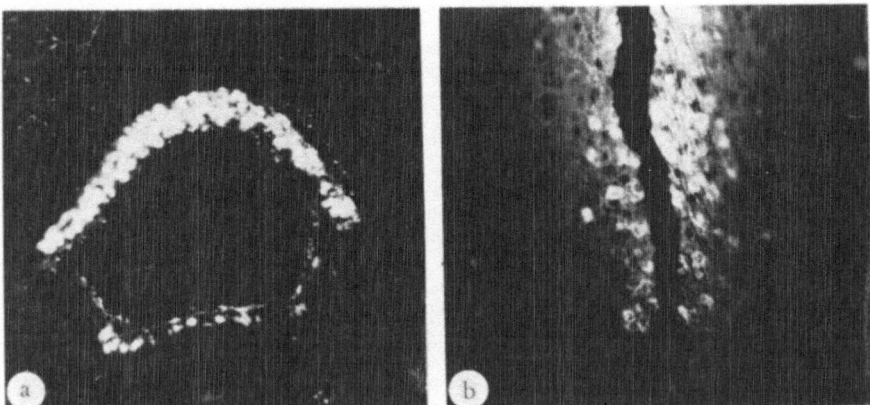

Fig. 16. Serotonin fluorescence in mouse treated with nialamide (250 mg/kg). Pineal gland (a) and intraneuronal serotonin-containing cells of the medial raphé mesencephalic region (b). 120 ×. (Reduced for reproduction 20%.)

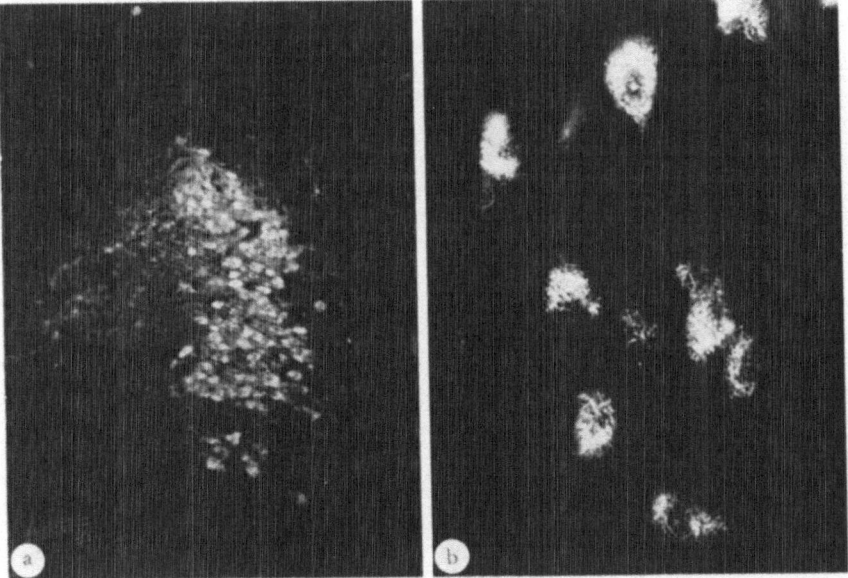

Fig. 17. Locus coeruleus in the rat (nialamide, 250 mg/kg). (a) Mass of catecholamine cell bodies. 40 ×. (b) Same as (a). 120 ×. (Reduced for reproduction 30%.)

warmed over a hotplate. Paraffin is dissolved by covering the section with an Entellan-xylene (Entellan, Merck) mixture (1:1 ratio), and a cover slip is placed over the section. (9) The slide may be rewarmed in a hotplate

of the monoamines sometimes varied in direct relationship to the time of death and to the enzymatic destruction and catabolism in nerve tissue. The histochemical fluorescence technique for the human brain, although far from perfect, provided a means of visualization of the nerve endings and neurons, localization of them in the brain, and semiquantitative determination of their concentration in these various regions. From eight human brains processed from 45 min to 1 hr 30 min after death, it was found that only tissue that had remained dead for not more 45 min was appropriate for histochemical analysis. Monoamine visualization in all brains taken 45 min after death was, without exception, worthless as far as catecholamine detection was concerned. The study reported here is therefore from only one brain out of three that gave relatively satisfactory localization of endogenous catecholamine fluorescence. An additional human brain was recently

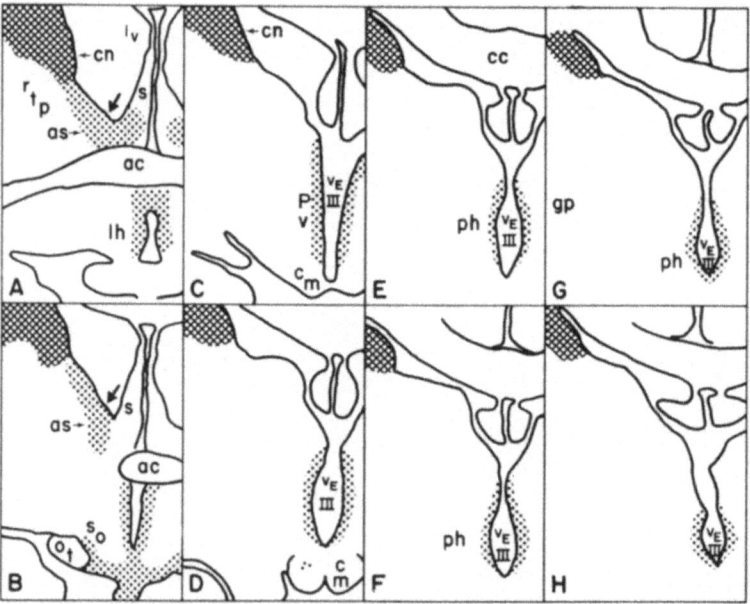

Fig. 19. Cross sections of human diencephalon and adjoining structures showing catecholamine-containing terminals and cell bodies (dots). Diffuse, terminal catecholamine fluorescence is seen (cross-hatching) in caudate sections (sections A–H, cn). Periventricular (V III) terminal fluorescence extends to supraoptic (section B, so) and lateral hypothalamic nuclei (section A, lh). Very few cell bodies are seen just dorsal to mammillary body (section D, cm). Heavy catecholamine fluorescence in the nucleus accumbens septi (arrow) extending from the base of the caudate to the septum (sections A and B) is seen as a confluent mass of synaptic terminals and neuronal cell bodies (see Fig. 20). Visualization of catecholamines by Falck–Hillarp technique. Magnification 40 ×. (Reduced for reproduction 30 %.)

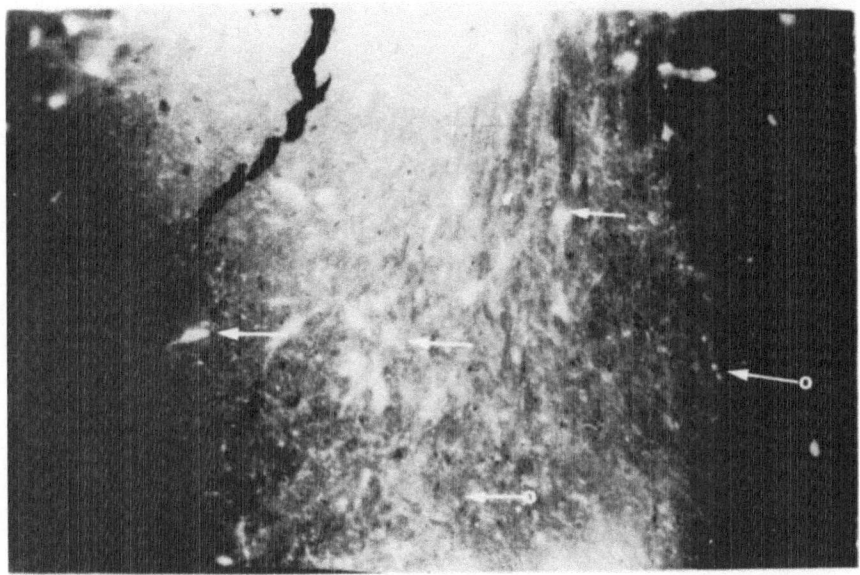

Fig. 20. Human brain, region of the nucleus accumbens septo. Heavy concentration of catecholamine-containing neurons (←) and synaptic terminals (←○). Diffuse fluorescence is seen at top from head of caudate nucleus. Falck–Hillarp technique. 120 ×. (Reduced for reproduction 45%.)

obtained (de la Torre, unpublished report) and processed in a new freeze-drying apparatus (Frystork FT-1, Bergman & Beving AB, Sweden); mono-amine localization (Figs. 19 and 20) was quite acceptable.

The brain was quickly removed at the autopsy room some 30 min after death and taken immediately to the laboratory, where subcortical sections were rapidly frozen in a precooled liquid mixture of isopentane–propane (50:50) to a temperature of −140 C. The temperature was further lowered to −170 C when the glass jar containing the tissues was immersed in liquid nitrogen. Lyophilization time was 10 days (see Section 2.4.1). After drying, the tissues were exposed to formaldehyde vapor at 50–70% relative humidity for 1 hr and then paraffinized, sectioned at 10 μ, and mounted in an Entellan–xylene solution on dry slides. All fluorescent preparations had a comparable hematoxylin and eosin section for identification of structures. Microscopic analysis of the human brain processed in this manner revealed a green, diffuse periventricular nucleus fluorescence in the rostrocaudal length of the hypothalamus. The specificity of catecholamine fluorescence was verified by omitting formaldehyde treatment and reducing the tissue with $NaBH_4$. Some scattered neurons containing a more intense fluorescence were noted against the periventricular background fluorescence (see

Chapter 3

Normal and Abnormal Physiology of Monoamines in the Central Nervous System

3.1. MONOAMINE UPTAKE AND DISTRIBUTION IN BRAIN

Noradrenaline and adrenaline were first detected in mammalian brain by von Euler (1946), and their presence was later confirmed by Holtz (1950). The NA that was found together with A was referred to as "sympathin" and was thought to occur in cerebral vasomotor nerves. In 1954, Vogt approached the problem of whether these amines have a functional role in the central nervous brain tissue, and in what is now a classic study in the field, she described the concentration and location of brain NA in several mammals. She found that in the dog the concentration of brain NA is usually uneven and does not parallel brain vascularity. Most of Vogt's results have been confirmed (see Table 4), with slight variation in concentration determinations due to technique.

The distributions of dopamine and NA in mammalian brain are markedly different. Dopamine is found to be highest in the neostriatum and substantia nigra, while NA tends to accumulate most in the hypothalamus and brain stem. Noradreanline is localized in all postsynaptic sympathetic neurons as well as in adrenergic nerve fibers and in the ganglia from which these fibers originate. The ability to take up NA is therefore lost in these sympathetic nerves after postganglionic sympathetic denervation. *In vivo* studies of NA uptake are difficult to perform due to the blood–brain barrier, but intraventricularly injected NA shows concentration of this amine *in vitro*. Uptake of NA may also be studied by a procedure using brain slices which have been gently homogenized to form synaptosomes (pinched-off nerve endings) in the vicinity of nerve terminals. These include mito-

Table 4.   Normal Concentration of NA and Dopamine in the Dog[a]

Structure	NA (Vogt, 1954)	NA (Carlsson, 1966)	Dopamine (Carlsson, 1966)
Hypothalamus	1.03	0.76	0.26
Midbrain	0.37	0.33	0.20
Thalamus	0.23	—	—
Cortex (cerebellum)	0.07	—	—
Cortex (cerebrum)	0.11	—	—
Cerebellum	—	0.06	0.03
Caudate nucleus	0.06	0.10	5.90
Pons	0.20	0.41	0.10
Medulla oblongata	0.45	0.37	0.13

[a] Results are expressed as $\mu g/g$ of tissue.

Table 5.   Occurrence of Catecholamines in Brains of Various Animal Species[a]

Species	Dopamine ($\mu g/g$)	NA ($\mu g/g$)	A ($\mu g/g$)
Dog	0.19	0.16	––
Cat	0.22	0.16	––
Frog and toad	—	0.26	1.40
Guinea pig	0.34	0.38	––
Pig	0.22	0.14	––
Rabbit	0.28	0.22	––
Rat	0.60	0.49	––
Sheep	0.30	0.25	––

[a] Note similarities in quantities of dopamine and NA. Data from Carlsson (1966).

chondria and amine-containing vesicles (see Section 2.3). About 30% of NA in bovine hypothalamus may be recovered in the isolated synaptosomes, and when osmotic shock of amine-containing vesicles is tried, NA may also be recovered in significant amounts by fractionation. Dopamine and NA are present in mammalian brain in roughly equal amounts in various species (see Table 5). The various organs to which NA and A are delivered decide the fate of these circulating amines with respect to storage and uptake. Organs such as the liver contain high levels of COMT, which transforms the delivered NA and A to their O-methylated metabolites normetadrenaline

The cellular fluorescence, which is localized in the cell cytoplasm, is evident in the reticular formation as a large group that accumulates just caudal to the nucleus ruber joining thin, varicose fluorescing fibers. They are also scattered laterally to the reticular formation and ventrally to the nucleus ruber tapering off in intensity after a cranial progression rising toward the middle level of the nucleus interpeduncularis.

Dopamine has also been detected histochemically in the neostriatum where a rather diffuse, strong fluorescence is observed. This diffusion may depend on the diffuse distribution of dopamine or may be due to the localization of the amine in submicroscopic structures interlaced and tightly concentrated to give the diffuse appearance.

In the substantia nigra, the catecholamine cell types (Fig. 22a, b, d) are found within the zona compacta and to a lesser extent within the zona reticulata. In this area, fiber bundles rise from Forel's field H_2 and from the ventral section of the crus cerebri which pass through the capsula interna and rise rostrally to end in the neostriatum or in the lateral part of the hypothalamus.

Dorsal to the nucleus interpeduncularis lies the largest group of CA cells which includes an area of cellular fluorescence within the nucleus Edinger–Westphal, nucleus linearis (pars intermedialis), and the nucleus of the oculomotor nerve.

Serotonergic nerve cells are found concentrated in the substantia grisea centralis while the majority of 5-HT nerve cells are seen in the nucleus dorsalis raphé (Fig. 22c). Above and medial to the medial forebrain bundle there are a number of cells that extend laterally to just behind the nucleus Edinger–Westphal. Cells containing 5-HT are also present in the caudal section of the nucleus linearis and the nucleus medianus raphé which enter the reticular formation. Yellow cellular fluorescence is evident around the lemniscus medialis and at the mesencephalo–diencephalic junction where they appear concentrated in an area dorsomedial to the fasciculus retro-flexus passing through the dorsolateral part of the mammillary body where the fibers seem to end.

In the diencephalon, catecholamine green fluorescence appears around the third ventricle and around the fasciculus retroflexus mostly within the substantia grisea periventricularis. The cells run cranially within the posterior hypothalamic nucleus, supramammillary area, and the nucleus reuniens thalami. There is a weaker green fluorescence in the anterior part of the nucleus arcuatus just above the median eminence which in contrast exhibits a strong fluorescence of multiple nerve fibers.

The location of 5-HT-containing cells in the central gray matter of the midbrain coincides with a region where cells are found to respond in a characteristic fashion to various sensory stimuli in the rabbit. These cells may function in part as a *startle-response* mechanism which occurs when the

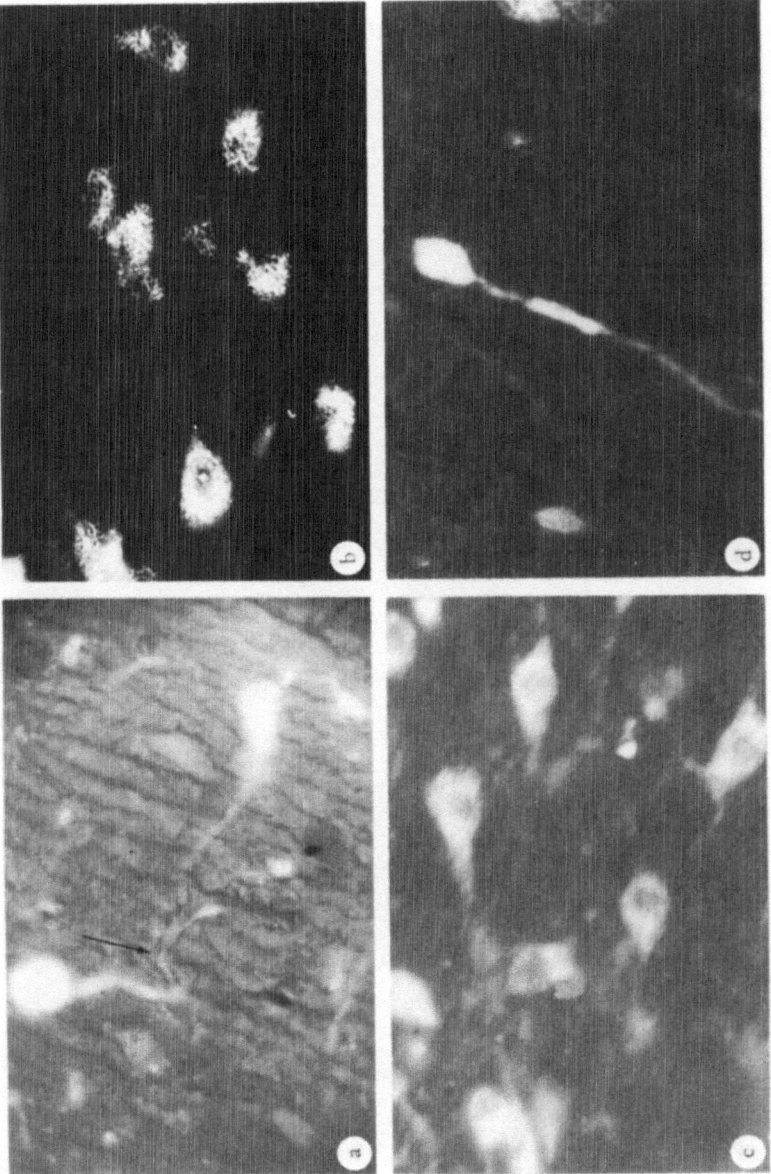

Fig. 22. Central monoaminergic neurons in rat brain. (a) Two catecholamine-containing cell bodies making contact with a capillary (arrow) in the substantia nigra. Increase in neuronal fluorescence is due to penetration of L-dopa after prior treatment with MK-486. 180 ×. (b) Catecholamine-containing nerve cells (normal) from substantia nigra. 400 ×. (c) Serotonin-containing nerve cells from dorsal raphé after prior treatment with nialamide. 400 ×. (d) Catecholamine-containing neuron and axon from substantia nigra after nialamide. 120 ×. (Parts a–d reduced for reproduction 30%.)

Table 6. Nerve Terminal Fluorescence in Brain Following Monoamine Oxidase Inhibition by Parenteral Nialamide Administration in Rats[a]

Structure	CA terminals	5-HT terminals
TELENCEPHALON		
(a) neocortex	⊙	⊙
(b) neostriatum	●	○
(c) globus pallidus	○	⊙
Amygdaloid complex		
(a) lateral nucleus	⊙	○
(b) central nucleus	●	⊙
(c) anterior nucleus	⊙	⊙
(d) cortical nucleus	⊙	⊙
Septal region		
(a) lateral nucleus	◑	⊙
(b) medial nucleus	⊙	○
(c) nucleus accumbens	●	○
(d) stria terminalis	●	○
(e) diagonal band	⊙	⊙
(f) olfactory tubercle	⊙	○
Hippocampus	●	⊙
Entorhinal cortex	⊙	⊙
Cingulate cortex	⊙	⊙
Pyriform cortex	◐	⊙
DIENCEPHALON		
Hypothalamus		
(a) ventromedial nucleus	⊙	○
(b) dorsomedial nucleus	●	⊙
(c) arcuate nucleus	◑	○
(d) tuberoinfundibulum	●	⊙
(e) supraoptic nucleus	●	○
(f) suprachiasmatic nucleus	○	●
(g) anterior nucleus	⊙	⊙
(h) preoptic area (medial)	◑	⊙
(i) preoptic area (lateral)	●	○
(j) lateral hypothalamic nucleus	◑	○
(k) periventricular nucleus	●	○
(l) paraventricular nucleus	●	○
Mammillary region		
(a) premammillary area	◑	⊙
(b) medial mammillary nucleus	⊙	⊙
(c) lateral mammillary nucleus	⊙	⊙
(d) posterior mammillary nucleus	◑	⊙
Epithalamus		
(a) medial habenula	⊙	⊙
(b) lateral habenula	⊙	○

Table 6 (Cont'd)

Structure	CA terminals	5-HT terminals
Subthalamus		
(a) subthalamic nucleus	○	○
(b) reticular nucleus	◉	○
(c) zona incerta	◉	○
Thalamus		
(a) anteroventral nucleus	●	◉
(b) anterodorsal nucleus	◉	○
(c) medial geniculate	◉	◉
(d) lateral geniculate	◑	◉
(e) ventral nucleus	◉	○
(f) lateral posterior nucleus	◉	◉
MIDBRAIN		
(a) substantia nigra (pars compacta)	◉	○
(b) substantia nigra (pars reticulata)	○	○
(c) oculomotor nerve	○	○
(d) Edinger–Westphal nucleus	◉	◉
(e) reticular formation	●	◉
(f) superior colliculus	◉	◉
(g) inferior colliculus	◉	◉
(h) pretectum	◉	◉
(i) periventricular gray (ventromedial)	◑	○
(j) periventricular gray (ventrolateral)	●	○
(k) red nucleus	○	○
CEREBELLUM	◑	○
PONS–MEDULLA		
Raphé complex		
(a) raphé obscurus	◑	◉
(b) raphé pallidus	●	◉
(c) raphé magnus	◑	◉
(d) raphé dorsalis	◑	◉
Carnial nerve nuclei		
(a) hypoglossal	◑	◉
(b) trigeminal (sensory)	◉	◉
(c) trigeminal (motor)	◑	◑
(d) trigeminal (mesencephalic)	◉	○
(e) abducens	○	○
(f) cochlear	◉	○
(g) ambiguus	◉	◉
(h) dorsal vagus	●	●
(i) solitarius	●	●
(j) vestibular	○	○
Inferior olive		
(a) dorsal part	●	○
(b) vental part	◉	◉
(c) dorsal accessory	○	◉
(d) ventral accessory	◑	◉

These terminals are seen more concentrated in the areas of the ventro-medial hypothalamic nucleus (Fig. 23a), dorsomedial hypothalamic nucleus (Fig. 23b), and surrounding the anterior commissure (Fig. 23c). Nor-adrenergic terminals are also typically found in close proximity to the length of the third ventricle (periventricular). The supraoptic nucleus contains many catecholamine-containing nerve terminals just above the optic tract. Noradrenergic sympathetic innervation of blood vessels may be seen to fluoresce mainly in the adventitial layer of arteries (Fig. 23d) and arterioles, but not in venules or capillaries. Although fluorescent fibers have at times been reported innervating veins in the choroid plexus, this type of innervation is quite rare. A more detailed discussion of noradrenergic fibers in vessels follows in the section discussing vascular mechanisms (3.2.3). Dopamine and noradrenaline terminals have been described by Bjorklund *et al.* (1970) in the rat median eminence by means of micro-spectrofluorimetric technique. More recently, Brown and Hornykiewicz (1971) reported that median eminence noradrenaline is present in higher concentrations than dopamine as measured biochemically. This would support the contention that tubero-hypophyseal monoamine fiber systems may be involved in anterior pituitary hormonal secretions (see also Section 4.3.1).

3.4. MONOAMINES AS TRANSMITTERS IN THE CENTRAL NERVOUS SYSTEM

It is generally accepted that neurotransmitters in the central nervous system function to modulate or initiate a sequence of reactions with respect to synaptic structures and neuromuscular junctions. *Subsynaptic receptor* is a term applied to the area of the postsynaptic membrane that is receptive to the chemical transmitter. The problem of studying receptor sites, aside from its extraordinary complexity, is compounded by the fact that this structure has not yet been isolated. Thus, only indirect approaches are available to the investigator for gleaning information on drug–receptor interactions. Two methods which have yielded substantial results are (1) physiological or chemical blockade of the transmitter and (2) interaction of cell receptors with various antagonists. The reader is referred to the excellent monographs by Eccles (1964) and McLennan (1969) on this subject.

Evidence is accumulating to suggest a central role for the monoamines in synaptic transmission function. This evidence can be summarized as follows: (1) localization of monoamines in neuronal cell bodies and nerve terminals by use of the histochemical fluorescence technique, (2) the finding of synaptic vesicles similar to those found in peripheral adrenergic nerve endings, (3) the finding of regional differences between monoamines and their enzymes in the central nervous system, (4) discovery of the presence of elevated catecholamine concentrations in fractions of brain homogenates that are rich in synaptic vesicles, and (5) results of pharmacological studies using agents administered by microelectrophoresis.

Thanks in part to the microelectrophoretic technique, studies have dealt with the response of NA to nerve cells in various brain structures. The general picture that emerges from these investigations is that in all brain regions examined there are usually some neurons sensitive to NA administration. Earlier reports on the absence of NA-sensitive neurons in the lumbar segments of the spinal cord and the medulla oblongata have been put in doubt by more recent investigations using monoamines administered by microelectrophoresis. In addition to NA-sensitive neurons, dopamine-sensitive neurons are also found. Results with this technique then, coupled with other pharmacological and histochemical evidence, suggest the presence of monoaminergic synapses in the mammalian central nervous system. Moreover, neurons having this type of synapse would be expected to show a response to locally administered monoamine, while other neurons lacking this synapse would remain insensitive. NA is known to produce either of two effects in a sensitive neuron, depression or facilitation of neuronal activity. These two elemental mechanisms underlie all function of the central nervous system and can modify transmission signals or modulate the activity of nerve cells. As regards synaptic transmitters in the central nervous system, McLennan (1967) has put forward three fundamental requirements for a transmitter molecule: (1) release—the transmitter molecule must be released from the nerve terminal upon nerve activity; (2) mimicry—the administered molecule will mimic the effect of synaptic stimulation upon direct application; and (3) pharmacology—if a pharmacological agent is known to interfere with the synaptic process, then it should also interfere with the exogenously applied compound.

In the putamen, dopamine can be released by direct stimulation as well as by stimulation of the substantia nigra. This would seem to conform to the hypothesis of Andén et al. (1966a) of a dopaminergic pathway between the substantia nigra and the neostriatum. The pathway may also be blocked to dopamine transmission pharmacologically by use of reserpine, haloperidol, or chlorpromazine or physically by unilateral lesions in the crus cerebri. It has also been shown by Ungerstedt et al. (1968) that unilateral intrastriatal injection of dopamine and NA in nialamide-pretreated rats will produce asymmetrical posture and turning contralateral to the site of the injection. This effect can be blocked by intrastriatal injection of chlorpromazine. This finding adds to the growing evidence of a role for dopamine in certain types of postural and locomotor activities.

Facilitation and nerve cell depression have been studied in the hypothalamus, in the caudate nucleus, and in the spinal cord. Neurons of the spinal cord in the lumbosacral segment were found to be either excited or depressed on administration of microquantities of NA and 5-HT. Most Renshaw cells were depressed by NA, but occasionally a few showed excitement. Weight and Salmoiraghi (1967) believe that Renshaw cells may belong to at least two different groups in spite of evidence that they all respond to ventral root stimulus. A variety of responses to NA, 5-HT, and ACh have been obtained with interneurons. Some of these cells responded to ACh but not to NA or 5-HT; others responded to NA but not to ACh or 5-HT; still others responded to 5-HT but not to ACh or NA. By contrast, none of the γ-motoneurons (which innervate muscular fibers) examined showed any facilitation of firing during microelectrophoretic administration of NA. Dahlström and Fuxe (1965a), using the histochemical fluorescence criteria, studied the retrograde degeneration of spinal cord neurons to find out whether or not the cell bodies of γ-motoneurons receive 5-HT and NA terminals. They found a close contact between many processes and cell bodies of γ-motoneurons and NA and 5-HT terminals. However, there exists the possibility that the monoamine terminals observed may make synapses with nonmonoaminergic fibers which in turn form the synapses with the motoneurons and their processes. In the ventral horn of the spinal cord, there are 5-HT and NA terminals present in contact with smaller cells that could be γ-motoneurons (which innervate muscle spindle receptors) or interneurons or both.

There is also evidence presented by Salmoiraghi *et al.* (1964) that NA is a synaptic transmitter in an inhibitory pathway in the rabbit olfactory bulb. Within this structure lies an arrangement of concentric layers of various nerve cell types: glomerular, tufted, mitral, and granular. The axons of mitral cells make up the lateral olfactory tract, which ends up in the prepyriform cortex. Electrophoretic administration of NA produced a response in the mitral cells; further biochemical analysis disclosed the presence of NA but not of A or dopamine in the olfactory bulb of untreated animals.

There appears, therefore, to be strong evidence that Renshaw cells as well as many other types of cells in the spinal cord are responsive to NA and 5-HT. In the medulla oblongata, there is substantiated evidence of the presence of neurons which respond to ACh, 5-HT, and NA. Histochemical data from the monoamine-containing fibers in the hypothalamus seem to correlate with the distribution of NA-sensitive cells in that structure. Cell fractionation and ultramicroscopic results appear to complement the data of monoamine distribution in the various submitochondrial fractions isolated from brain nerve endings. The sum total of these findings indicates a neurotransmitter role for dopamine, NA, and 5-HT in certain regions of the central nervous system.

3.4.1. Monoaminergic Pathways

Several projections of monoaminergic fiber tracts have been described so far in the rat by use of the fluorescence technique. One group of cells gives off a fiber tract which originates at the substantia nigra and ends with dopamine-containing terminals in the neostriatum.

Dahlström and Fuxe (1964a) have shown that dopamine-containing nerve cells are present in the pars compacta of the substantia nigra and also in the neostriatum which appear localized as fine or submicroscopic catecholamine terminals. If electrolytic lesions are made in the substantia nigra or internal capsule, the histochemical fluorescence and the dopamine content of the neostriatum are markedly reduced, while removal of the neostriatum produces an increase in fluorescence of the dopaminergic cells in the substantia nigra and in their axons central to the lesion. Moreover, when a large part of the neostriatum is removed unilaterally, most of the catecholamine-containing cell bodies within the substantia nigra— but nowhere else—appear to be increased in fluorescence intensity and swollen 3 or 4 days after ablation. After 3–4 weeks, the fluorescence intensity decreases, and degenerative changes are observed. Complementing the changes in these cell bodies is a rapid accumulation of catecholamines within the swollen nerve fibers in the internal capsule. The nerve fibers can be traced caudally through the retrolenticular part of the internal capsule down to the crus cerebri.

Another monoaminergic nerve tract has been described by Carlsson et al. (1964) which has smaller groups of cells in the medulla oblongata sending axons caudally to the spinal cord to end in NA and 5-HT terminals in the gray matter.

Noradrenaline cell bodies are found in the pontomedullary region; two small groups of these neurons have been described in the rubrospinal tract just lateral to the superior olivary nucleus. The second group is situated in the reticular formation of the pons slightly ventral to the superior cerebral peduncle. The largest group of noradrenaline-containing cell bodies is found in the locus coeruleus. These fibers may project anteriorly to join the medial forebrain bundle.

Two bulbospinal noradrenergic pathways have been described in the rat. One of these runs caudally into the lateral margin of the anterior horn and the ventral part of the lateral funiculus, while the other descends in the lateral funiculus and terminates in the sympathetic lateral column and in the posterior horn, particularly in the substantia gelatinosa (Fig. 24). Most terminals in the spinal cord, whether 5-HT or noradrenergic, appear to have similar distribution. The density of monoamine terminals in the cord can be noted by the rate of amine disappearance after selective blocking of their synthesis. Axonal serotonergic projections can best be visualized after

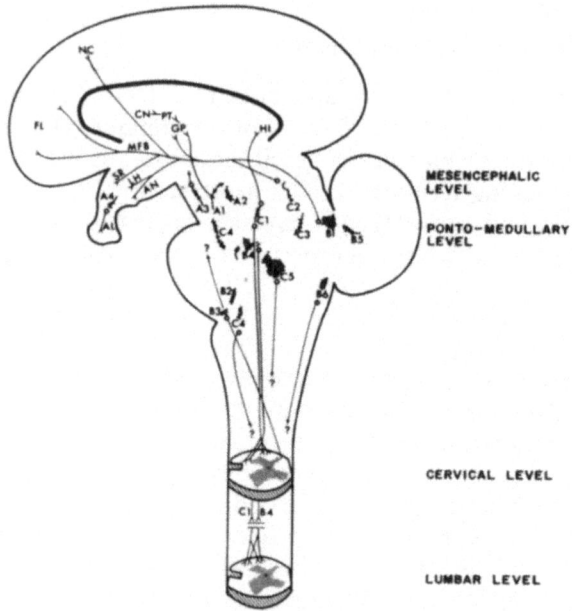

Fig. 24. Monoaminergic pathways. Schematic representation of presumed monoaminergic pathways in the human brain based on animal experimentation. AL, anterior lobe pituitary; AN, amygdaloid nucleus; CN, caudate nucleus; FL, frontal lobe; GP, globus pallidus; HI, hippocampus; LH, lateral hypothalamus; MFB, medial forebrain bundle; NC, neocortex; SR, septal region.

Origin of neuronal cell bodies	Afferents to	Presumed pathway
(A) Dopaminergic pathways		
(1) Substantia nigra	Striatum	Crus cerebri and lateral part of medial forebrain bundle (MFB)
(2) Rostral mesencephalic reticulum	?	?
(3) Interpeduncular nucleus	Rostral forebrain	?
(4) Tuberal region	Anterior hypophysis	Tuberoinfundibulum
(B) Noradrenergic pathways		
(1) Locus coeruleus	Forebrain	Medial forebrain bundle (MFB)
(2) Rubrospinal tract	Anterior horn (lateral part)	?
(3) Lateral aspect of superior olive	Substantia gelatinosa	Lateral funiculus
(4) Lateral medullary reticular formation	Anterior and lateral funiculi	?
(5) Ventral aspect of brachium conjunctivum	?	?
(6) Dorsal vagus nucleus	Spinal Cord	Bulbospinal
(C) Serotonergic pathways		

Origin of neuronal cell bodies	Afferents to	Presumed pathway
(1) Midline raphé region	Septal region	Medial forebrain bundle (MFB)
(2) Dorsal raphé nucleus	Lateral hypothalamus, globus pallidus, hippocampus	MFB
(3) Caudal median raphé	Neocortex, amygdaloid nucleus	MFB
(4) Area around medial lemniscus	Spinal cord	?
(5) Nucleus raphé pallidus (level of facial nerve)	Spinal cord	?

monoamine oxidase inhibition. Catecholamine-containing axons can be traced after axotomy by determining whether the fluorescence has increased proximally to the axon or diminished distally to the lesion. Neuronal pathways can be mapped out using histochemical fluorescence after nerve sectioning. Changes seen antegrade and retrograde to the lesion can then be followed on ascending and descending monoamine tracts. Ascending fiber tracts have been reported to arise from cells in the midline raphé, ventro-lateral reticular formation, and ventral part of the periventricular and central gray matter.

The medial forebrain bundle, which contains 5-HT, projects from cells in the raphé and also from the caudal end of the superior colliculus. Reports by Heller and Moore (1970) indicate that selective lesions lateral and ventral to the tegmentum would affect brain noradrenaline levels only, while medial destruction of the central gray matter would affect serotonin exclusively.

The remainder of 5-HT cells are located in the medulla, ventral arcuate nucleus, area postrema, nucleus obscurus, raphé pallidus, and the region surrounding the pyramidal tracts at the level of the facial nerve nucleus.

3.4.2. Monoamines in Brain Embryonic Development

Another important method of relating function with neurochemical correlates of behavior in an organism is the study of monoamines and their enzyme systems in the developing brain. Thus Brodie et al. (1959) have proposed that NA and 5-HT may modulate primitive patterns of behavior in the maturing animal. Animals such as the rat and the rabbit develop their specific patterns of behavior at 2–3 weeks in their postnatal life, while the guinea pig is born with a well-developed central nervous system and exhibits marked behavioral patterns soon after birth.

Karki et al. (1962) have investigated the synthesis and metabolism of 5-HT and NA in the developing brains of rats, guinea pigs, and rabbits. There are considerable differences in the functional development of new-borns of these species, as revealed by variations in NA and 5-HT newborn–adult ratios (Table 7). Two days before birth, rats have a 5-HT level 30% less than the adult level, and there is no detectable NA. At birth, the level of 5-HT rises to 55% and NA to 20% of the adult level. These levels remain more or less the same during the first postpartum week and begin to rise in the second week, with adult concentrations of 5-HT and NA being reached after 3 weeks and 6 weeks, respectively. By contrast, brain levels of 5-HT and NA in newborn guinea pigs are almost identical to the adult levels. In young rabbits, the concentration of both amines is 50% of the adult level. The concentration of the decarboxylating enzyme of dopa and 5-HTP is about the same as the level in young guinea pigs and rat brains, while the monoamine oxidase activity is only 30% of the adult activity in young rats and about that percentage in young guinea pigs.

Although there are serious methodological limitations to attempting amine determinations in the developing brain, there appears to be a good correlation between monoamines present and the enzymes responsible for their synthesis. Schadé (1959) reported that between the eighteenth and twenty-first days of life 100% of the rabbits he studied showed an EEG adult pattern. McCaman and Aprison (1964) reported a 5-HTP–dopa decarboxylase activity of 50–70% of the adult level in rabbits 21 days old. These differences in results suggest that approach and the relative sensitivity of the various methods used in the determinations may be the sources of the differences. McCaman and Aprison (1964) observed in their studies an interesting caudal–rostral development of several enzymes and MAO activity in the young rabbit. MAO activity was noted 3 days after birth in the medulla oblongata extending gradually toward the cerebral cortex and caudate nucleus, where it was finally detected after 19 days.

Haber and Kamano (1966), using pooled homogenates of gradually maturing rat brains, determined that 1 day after birth the rat brain has

Table 7.   Concentrations of Norepinephrine NA and Serotinin 5-HT in the Brains of Adult and Newborn Animals[a]

Species	Adult		Newborn	
	NE (μg/g $\pm$ SD)	5–HT (μg/g $\pm$ SD)	NE (μg/g $\pm$ SD)	5–HT (μg/g $\pm$ SD)
Rat	0.54 $\pm$ 0.09	0.47 $\pm$ 0.02	0.12 $\pm$ 0.02	0.26 $\pm$ 0.02
Rabbit	0.28 $\pm$ 0.03	0.60 $\pm$ 0.02	0.14 $\pm$ 0.02	0.33 $\pm$ 0.05
Guinea pig	0.34 $\pm$ 0.01	0.33 $\pm$ 0.01	0.26 $\pm$ 0.01	0.26 $\pm$ 0.01

[a] Each value represents the mean obtained from at least six determinations. For each determination, ten newborn or one or two adult brains were pooled. From Karki et al. (1962).

39% of the adult 5-HT content but only 17% of the adult weight. They found the ratio of brain weight to serotonin in unity after 1 day post-natal. A gradual increase in brain weight/serotonin content follows for 10 days and thereafter remains constant until adult life indicating equal maturation rates for both wet brain weight and serotonin concentrations. These results confirm those of Karki *et al.* (1962), who also suggested that the immature rat brain is not able to bind 5-HT as efficiently as the adult brain. This deficiency of indoleamine binding in the newborn rat brain is not a reflection of an insufficient enzyme system, which is in fact present at an early age. The deficiency seems comparable to the discovery of Glowinski *et al.* (1964) that circulating NA cannot be readily bound in the developing heart. Bennett and Giarman (1965) found a major increase of 5-HT and MAO in rats after birth, whereas 5-HTP decarboxylase activity had already reached the adult level at birth. They also showed that newborn and adult rats synthesize cerebral 5-HT from blood-borne DL-5-HTP at equal rates but that the young rat's ability to synthesize 5-HT from blood-borne L-tryptophan is less. This indicates that the prime cause of the low 5-HT level in the young rat brain is its reduced capacity to hydroxylate tryptophan. Chronic administration of MAO inhibitors (α-methyldopa, NSD-1034) does not seem to alter the development of 5-HT metabolism in brains of fetal or neonatal rats.

Guroff and Udenfriend (1964) reported that tyrosine uptake in newborn animals is more pronounced than in adults. Their study also indicated that all aromatic amino acids share a common transport pathway and compete for entry into the brain. Even though results show a faster uptake of L-tyrosine in newborn rat brains, the stereospecificity and brain barrier function remain equally pronounced in the newborn and adult animals. Thus the blood–brain barrier remains unaffected in newborn or adult, but the transport mechanism in the newborn is more active for various ions and metabolites than its adult counterpart. This would explain why previous reports indicated an absence or reduction of blood barrier functions in the newborn even though the classical indicator of barrier function, trypan blue, was showed function even in fetal animals.

In cultures of human fetal brain, high concentrations of 5-HT appear to arrest oligodendrocytes in a contracted state. In the same culture, neither the microglial activity nor the protoplasmic astrocytes seem to be affected by the high content of serotonin except for nonspecific transient contractions.

The evidence from *in vitro* and *in vivo* studies of the developing mammalian brain indicates the importance of monoamine systems occurring in the maturing brain to subsequent adult behavioral and functional patterns.

3.4.3. Vascular Mechanisms and Catecholamines

It has long been known that the cerebral vessels are innervated by sympathetic fibers arising from the superior cervical ganglion. The sympathetic fibers extending toward the head join the plexus with the common carotid artery and are continuous with those of the internal and external carotid arteries. Other fibers join the vertebral artery, which is continuous with the plexus of the basilar artery. The internal carotid and the basilar arteries then join to form an anastomosis at the circle of Willis. The sympathetic fibers accompany blood vessels into the hypophysis. On the other hand, the external carotid plexus gives rise to subordinate plexuses on the branches of the external carotid artery, through which the fibers reach their peripheral distribution. This vascular innervation has recently been shown by histochemical fluorescence, which has clarified the two routes of efferent post-ganglionic fibers from the superior cervical ganglion. One route proceeds through the rostral postganglionic pathway and the other through the caudal postganglionic pathway. The rostral postganglionic innervates the internal carotid artery, and the caudal postganglionic innervates the vertebral artery. Thus the sympathetic innervation reaches the midline artery, the basilar artery, the anterior cerebral artery, and the posterior cerebellar artery. Postganglionic sympathetic nerves also supply the arteries leaving the circle of Willis. Histochemical fluorescence has also confirmed that after bilateral stellate ganglionectomy, the fluorescent innervation of the cerebral vessels is not affected (Kajikawa, 1969). This suggests that the vertebral arteries receive adrenergic nerve fibers only from the superior cervical ganglion. Superior cervical ganglionectomy produces total disappearance of the adrenergic fluorescence in the cerebral arteries. Little or no adrenergic innervation is seen in the pial veins. The fluorescent adrenergic innervation of the cerebral blood vessels is thought to contain noradrenaline. Cross sections of these vessels show that the noradrenaline is found mainly in the adventitial layer encircling the whole circumference of the artery (Fig. 25). No innervation is found in the venules or capillaries, but fluorescent fibers have been seen in small arterioles with diameters of about 15 μ.

The adrenergic innervation of arterioles forms the basis of vasotonic activity and the control of blood pressure in animal and in man. A typical contraction of the arteriole is thought to begin when the brain sends an electrical signal to the spinal cord along a nerve fiber which can stimulate acetylcholine at a junction. The acetylcholine in turn stimulates a second nerve fiber to transmit a signal that reaches the muscular wall of the vessel itself. At the muscle wall, noradrenaline is released, causing the arteriole to contract. On this basis, then, the vascular system can be controlled by drugs which either stimulate noradrenaline release at the muscular wall

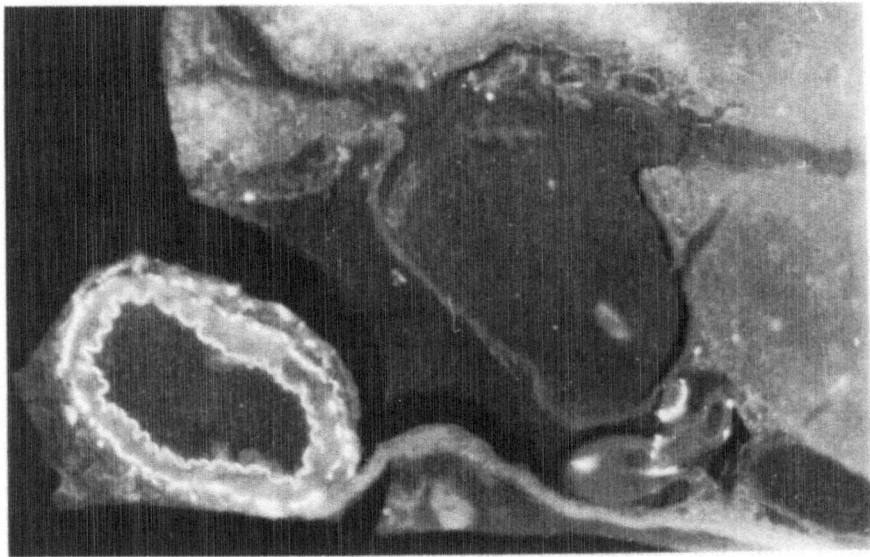

Fig. 25. Periarterial noradrenergic fluorescence in a pial vessel near the supraoptic nucleus of the rat. Autofluorescence (nonaminergic) is characteristically also seen in the elastica. 180 ×. (Reduced for reproduction 30%.)

of the vessel or, inversely, block its release. The so-called pressor drugs, such as adrenaline, can achieve the first response and raise the blood pressure. Other drugs which can block the original signal from the brain to the nerves, such as sedatives, or which can reduce the amount of noradrenaline release at the nerve endings, such as guanethidine, or which can prevent noradrenaline synthesis, such as α-methyldopa or disulfiram, can also lower the blood pressure. It is evident therefore that noradrenaline and possibly dopamine play an important role in the modulation of blood pressure and postural responses. A report by Barbeau (1970) linking dopamine and the renin–angiotensin system shows that an important factor modulating the sympathetic nervous system stimulation of renin secretion is the dopamine–noradrenaline ratio (DA/NA). Barbeau also pointed out that circadian rhythm may be related to dopamine excretion and that dopamine-rich structures, such as the basal ganglia, may be involved in the peripheral modulation of renin. Various hypertensive agents that can affect dopamine metabolism in the caudate nucleus were able to increase the turnover of dopamine as reflected by homovanillic acid, and these drugs appear to limit the synthesis of noradrenaline and subsequently the sympathetic outflow from the brain stem pressor centers. The conclusion of this study is that the basal ganglia may, through the rapid feedback regulation of dopamine turnover, control the modulation of sympathetic tone through the effect of dopamine on noradrenaline synthesis. If confirmed, this study could lead to other investigations on the role of dopamine in blood pressure adaptation and to the formulation of drugs which would stimulate the renin–angiotensin system.

Another method of studying the hypotensive effects of drugs has been through pretreatment with dopa decarboxylase inhibitors such as [N-(DL-seryl)-N'-2,3,4-trihydroxybenzyl] hydrazine (Ro 4-4602) and α-methyldopa hydrazine (MK-485). The former inhibitor in doses exceeding 200 mg/kg was given to rats followed by α-methyldopa, a hypotensive agent that limits the synthesis of noradrenaline in brain. It was found that the hypotensive effect of α-methyldopa was completely inhibited by pretreatment with Ro 4-4602 concomitant with a reduction in the accumulation of α-methyl-dopamine in the heart and brain. In addition, Ro 4-4602 also prevented the depletion of brain dopamine and of heart noradrenaline. If, however, the peripheral decarboxylase inhibitor MK-485 was given to rats before treatment with α-methyldopa (at the same dose), the hypertensive response was still evident but the accumulation of α-methyldopamine was inhibited as well as the depletion of noradrenaline in the heart. This dual drug combination, however, does not prevent the accumulation of α-methyldopamine or the depletion of dopamine and noradrenaline in the brain (Henning, 1969). These results show that, in the first case, the dose of the decarboxylase inhibitor Ro 4-4602 is extremely important to the outcome of the response. High doses of Ro 4-4602 can inhibit both peripheral and endogenous dopa decarboxylase activity in the brain with the result that less dopamine is available endogenously as well as less noradrenaline. With less dopamine available, a reduction in the methylation of the injected α-methyldopa results in less accumulation of α-methyldopamine. The inhibition of the hypotensive effects usually seen after α-methyldopa alone can be explained by the fact that Ro 4-4602 (at 200 mg/kg) has decreased the availability of dopamine, the precursor of noradrenaline, by inhibiting the transformation of dopa to dopamine endogenously. On the other hand, if MK-485 is used, and this agent only inhibits dopa decarboxylase activity peripherally, the endogenous dopa can proceed through its conversion to dopamine and to noradrenaline in a normal manner, and the follow-up injection of α-methyl-dopa will exert its normal effect on noradrenaline synthesis to reduce this amine in the brain and thus produce the hypotensive effect. This also explains why MK-485 does not deplete dopamine or noradrenaline from the brain or prevent the accumulation of α-methyldopamine in this organ as does the endogenous inhibitor Ro 4-4602 (at 200 mg/kg).

L-Dopa appears to have different effects on the vascular responses of man and animals. L-Dopa has been used mainly in parkinsonism patients, and the effect of this drug alone in high doses is usually to orthostatic hypotension which can only be relieved by α-adrenergic blockers such as methoxyamine. In animals, however, the opposite effect occurs. An injection of L-dopa (50 mg/kg) causes a rapid increase in the mean arterial blood pressure. The increase remains constant for at least 1 hr. Henning and Rubenson (1970) showed that injections of MK-485 do not influence blood pressure when given alone, but pretreating an animal with L-dopa can reverse the hypotensive effect of this amino acid. When a dopamine-β-hydroxylase inhibitor, FLA-63 (an isomer of tetraethylthiouram disulfide), was given to animals subsequently injected with MK-485 and L-dopa, the falling blood pressure seen after MK-485 plus L-dopa was abolished. Since it may be assumed that FLA-63 prevented the synthesis of noradrenaline from dopamine, Henning and Rubenson concluded that the central effect seen in this vasomotor reaction was probably linked to noradrenergic mechanisms rather than to dopamine. However, another interpretation is also possible. Pretreatment with M-485 followed by L-dopa probably abolishes the hypertensive effect of L-dopa alone because the latter is able to penetrate the blood–brain barrier into the brain substance, where it is subsequently taken up by the nerve cells and decarboxylated to dopamine endogenously. We have already discussed how other psychoactive agents which block or inhibit the synthesis of noradrenaline also lower the blood pressure. One can speculate, therefore, that if FLA-63 does not by itself have any effect on the blood pressure, as was reported by Henning and Rubenson, then its action may be other than dopamine-β-hydroxylase inhibition. If dopamine-β-hydroxylase inhibition by FLA-63 does indeed occur and the fall in blood pressure after MK-485 plus L-dopa is abolished, then it would be logical to assume that FLA-63 reverses the effects of MK-485 by permitting unusual accumulation of dopamine in the brain cells while these in turn exert some pressor effects on the vascular system. This would then make obvious the importance of dopamine accumulation in endogenous brain regions rather than noradrenergic mediation of these vasomotor responses. However, such problems must await further evidence to be solved conclusively, but it appears that when the overall picture of these investigations is made clear, it will be evident that while noradrenergic mechanisms are important in their innervation of cerebral vessels from sympathetic fibers, other chemical transmitters such as dopamine may also modulate sympathetic tone through some mechanism or series of mechanisms.

3.5. NEUROPHARMACOLOGY AND BEHAVIOR

The number of studies dealing with the interactions of monoamines with other drugs has been steadily increasing within the last few years.

Drugs that affect the monoamine balance in the brain have been the subject of much of this work. There are five kinds of drugs that can affect mono-amines in the central nervous system: (1) drugs acting specifically on the catecholamines, (2) drugs acting specifically on serotonin, (3) drugs acting specifically on either dopamine, noradrenaline, or adrenaline, (4) drugs acting on catecholamines and reversibly on serotonin either way, and (5) drugs acting similarly on monoamines in general. There may be variations, but these are the effects that yield the most valuable information regarding the monoamine balance in the brain. Histochemical fluorescence has become one of the most important tools in directly studying drug action on mono-amine nerve terminals and neurons. The advantage of this technique over biochemical assays is that specific amine metabolism in these structures may be visually evident. By contrast, with analytical assays the tissue is homo-genized or pooled without regard to structure. Thus, in a brain sample, blood vessels containing monoamines may be crushed together with central nervous tissue, resulting in inaccurate reflection of neuronal monoamine changes. Autoradiographic labeling and determination of ultrastructural changes in monoamine metabolism, while useful ancillary tools, have serious technical limitations with respect to drug–amine interactions.

In general, drugs which affect monoamine balance in the brain result in some behavioral changes in the living organism. These changes may be transient or irreversible. Guanethidine, which is an adrenergic blocking drug used in the treatment of hypertension, has been shown by Cox and Maickel (1969) to affect not only brain noradrenaline but also serotonin. In animals injected intraventricularly, guanethidine produced hyper-excitation and increased response to external stimuli within an hour. After 1 hr, prolonged depression replaced the hyperexcitability, mimicking the syndrome seen after reserpine, i.e., hunchback posture, ptosis, and ady-namia. Reserpine, of course, is known to produce a state of relaxation bordering on sleep, and yet it does not produce anesthesia. After reserpine treatment, the subject may be aroused by external stimulus, but the tranquil state returns immediately after cessation of the stimulation.

Another line of investigation is the study of the action of monoamines themselves administered to the organism. The Spanish endocrinologist Gregorio Marañon (1924) was one of the first to study the reaction to injected adrenaline of normal patients. The subjects reported feeling symp-toms of anxiety, fear, and anger, but the emotions were not of genuine character. When Marañon asked the patients to describe their feelings, they generally responded that they felt "as if" angered or anxious. Marañon later explained these sensations as "cold" emotions, not integral to the cognitive behavior pattern. In another series of experiments on normal volunteers, Ax (1953) measured the physiological responses after noradrena-line and adrenaline injections. Noradrenaline by itself produced symptoms resembling anger, while injections of combined noradrenaline and adrenaline resulted in a state of fear or anxiety. Since both noradrenaline and adrenaline do not penetrate the blood–brain barrier upon injection, these emotional reactions have been attributed in part to the interaction of the drugs with peripheral receptors. Another possibility may be that even minute quantities of either catecholamine that may pass through the brain barrier are sufficient amounts to alter or stimulate central receptors. Labeled adrenaline injected intravenously in animals can be recovered in barely measurable amounts in brain tissue, particularly the hypothalamus. The fact that traces of this amine can be recovered in the hypothalamus, a region implicated in various emotional states, may be evidence of the powerful activity exerted by adrenaline on behavioral reactions.

One way of circumventing the blood–brain barrier to monoamines is to give their precursors, 5-HTP for serotonin and dopa for dopamine and noradrenaline. It is easier to increase brain monoamine levels by pretreatment with a dopa decarboxylase inhibitor, but large doses of either precursor also increase endogenous monoamine concentrations. An indirect precursor of serotonin is tryptophan, which after enzymatic hydroxylation becomes the immediate serotonin precursor 5-HTP. Tryptophan given to normal humans produces states of well-being and drowsiness. Conversely, symptoms of depression may be treated with either a MAO inhibitor and dopa or large doses of dopa alone. L-dopa, which is currently being tested in humans afflicted with Parkinson's disease, is also reported to cause sexual stimulation after prolonged treatment. Another compound with this characteristic effect on the sexual behavior of rats is the serotonin inhibitor p-chlorophenylalanine (p-CPA). A p-CPA dose of 100 mg/kg for several days is reported by Tagliamonte et al. (1969) to increase copulatory and mounting behavior in rats. At these doses, p-CPA is capable of reducing serotonin to 20% of control values by inhibiting tryptophan hydroxylase, the rate-limiting step in the synthesis of brain serotonin. The decrease in catecholamines after p-CPA is less marked but nevertheless evident. Sexual arousal after manipulation of serotonin and noradrenaline is, however, still a subject of controversy. Additional investigation into the role of these amines in sexual behavior is necessary before any theory is advanced as to the functional causality of the so-called sex drive. Moreover, extrapolation of animal behavioral reactions to human emotion is difficult, if not impossible.

The most widely used drug in the world, alcohol, has been shown by Davis and Walsh (1970) to interact with the monoamines. Ethanol ingestion, this group reports, can inhibit normal oxidative catabolism of the intermediate aldehyde derivatives of serotonin and noradrenaline. The amine metabolism can also be diverted by alcohol to a secondary collateral reductive pathway, thus changing the normal metabolic reaction. Cross effects between indoleamines and catecholamines have also been observed. Moir (1969) reports that L-tryptophan can act to change dopamine synthesis in body fluids. Dopamine, on the other hand, may be involved in the pituitary regulation of various body hormones. For example, the luteinizing hormone necessary in ovulation can be blocked for a short time by selective inhibition of dopamine metabolism in the rat. Intracerebral injections of serotonin in guinea pigs may alter corticosteroid blood levels, suggesting a role for this amine in hypothalamic–pituitary–adrenal function.

The administration L-dopa is reported by Reis *et al.* (1970) to producing "sham rage" in cats. When cats were pretreated with a monoamine oxidase inhibitor followed by 20 mg/kg of L-dopa, intense excitement and sham rage dominated by fear behavior were seen. In contrast, Scheckel *et al.* (1969) reported sedation in rats after large doses of L-dopa (Fig. 26). In a study of drug-induced aggression in mice, Lycke *et al.* (1969) found that aggressive behavior followed increased dopamine synthesis provided that serotonin synthesis was also reduced.

In affective states, lithium is used as an antidepressant. Unlike many other drugs used for manic–depressive states, however, it does not seem to affect brain monoamine levels at all. But when a tryptophan hydroxylase inhibitor (α-propyldopacetamide) is used and then lithium, a lesser decrease in the depletion of serotonin from nerve terminals is observed. This may be due, according to Corrodi *et al.* (1969), to a lowering of activity in serotonergic neurons by chronic administration of lithium. Methysergide, which antagonizes serotonin in tissues and has a structure similar to that of LSD-25, has been used by Fieve *et al.* (1969) in the treatment of mania. This group reported little or no improvement in manic behavior after 7 days in six patients receiving methysergide but dramatic improvement was seen after lithium was substituted for methysergide. Increase in dopamine and concomitant decrease in 5-HT may be produced by L-dopa administration or induction of viral encephalitis for the dopamine increase and administration of p-CPA or α-propyldopacetamide for the 5-HT decrease. A number of psychotomimetics also characteristically influence monoamine metabolism as well as alter behavioral reactions, and this will be treated at more length in Section 4.2.

Another important aspect in studying monoamine changes in the brain is the use of enzyme inhibitors which interfere with normal amine metabolism. Characteristic of this group of compounds is N-(DL-seryl)-N'-2,3,4-trihydroxybenzyl] hydrazine (Ro 4-4602), which acts as a potent inhibitor of 5-HTP-dopa decarboxylase. Studies of this coumpound suggest that Ro 4-4602 may interfere mainly with the metabolism of free (unbound) amines, with slight action on stored amines. Ro 4-4602, which belongs to the hydrazine group, is capable of decreasing 5-HT and NA in brain by a maximum of 50% without affecting the MAO enzyme. The hydrazine differs

Fig. 26. Three possible reactions after L-dopa treatment with and without enzyme inhibitors in the rat. (1) Nialamide (MAOI) plus Ro 4-4602 (peripheral dopa decarboxylase inhibitor) + L-dopa. Animal shows no gross abnormal symptoms. (2) Nialamide plus L-dopa. Severe autonomic reactions, including hyperexcitability, exophthalmos, piloerection, salivation, tachycardia, hyperkinesia. (3) L-Dopa. Depression, apathy. Doses: Niallamide 25 mg/kg, Ro 4-4602 16 mg/kg, L-dopa 150 mg/kg.

from reserpine in that it decreases both serotonin and 5-hydroxyindoleacetic acid (5-HIAA). By contrast, administration of reserpine diminishes only 5-HT, with a parallel increase in 5-HIAA excretion. In the case of α-methyldopa, a similar decrease in both 5-HT and the 5-HIAA metabolite occurs, but the 5-HIAA excretion persists longer after α-methyldopa than after Ro 4-4602. A possible explanation may be that α-methyldopa affects the amines by a mechanism other than decarboxylase inhibition.

Another strong decarboxylase inhibitor is N-(3-hydroxybenzyl)-N'-methylhydrazine (NSD-1034), which is reported to block 75% of 5-HTP–dopa decarboxylase *in vivo*. NSD-1034 does not, however, produce more than a slight fall in brain 5-HT and dopamine, which continue to form even after MAO blockade. Tyrosine hydroxylase, which forms dopa in the metabolism of catecholamines, can be inhibited by α-methyl-*m*-tyrosine and α-methyltyrosine, resulting in a decrease in brain dopamine and NA without alteration of 5-HT levels. Spector *et al.* (1965) found that sedation could occur in animals after lowering brain NA with α-methyltyrosine. Brodie *et al.* (1966*a*) have confirmed this finding, but Koe and Weissman (1966*b*), in a parallel study, did not observe sedation. Conditioned avoidance response in animals also disappears after α-methyltyrosine, and it is probable that this behavioral response is more dependent on the reduction of brain NA than on concentration of the drug in the brain.

Depletion of tissue NA has been tried using disulfiram, which lowers the content of endogenous brain NA by inhibiting dopamine-β-hydroxylase, and it has been shown to be more effective in this sense than α-methyltyrosine. Furthermore, the decrease in NA is exceeded by the increase in accumulated dopamine in all brain regions analyzed. Goldstein and Nakajima (1967) have shown that only in animals treated with a MAO inhibitor is the decrease–increase ratio of NA–dopamine about equal in the hypothalamus and in the brain stem. Another more recent inhibitor of dopamine-β-hydroxylase, reported by Johnson *et al.* (1970), is 1-phenyl-3-(2-thiazolyl)-2-thiourea (U-14, 624), a specific depletor of NA *in vivo*. In mice and rats, a dose of 200 mg/kg of this enzyme inhibitor reduced NA to 10% of control values after 18 hr with a slight rise in dopamine levels. Should these findings be confirmed, another important drug will be available for the study of central monoamine functions. Hydroxamate derivatives have been reported by Utley (1966) to reduce dopa decarboxylase activity shortly after the hydroxames enter the brain, with catecholamine levels returning to normal as hydroxames are eliminated. At a dose of 1 mg/g, anthranilhydroxamic acid lowered brain catecholamine levels in mice less than salicylhydroxamic acid at 0.8 mg/g. Benzohydroxamic acid at 0.8 mg/g lowered not only catecholamine concentrations but also 5-HT content in the brain.

A neurotropic agent with apparent brain selective action in the cat is neuroketone (NK), prepared from ox brain as a complex of ketosteroidal substances. NK, when administered to cats, selectively increases serotonin levels in the hypothalamus only, while decreasing 5-HT content in the serum. A possible explanation is that NK is a serotonin antagonist capable of penetrating a weaker hypothalamic blood–brain barrier to exert its influence on the surrounding tissue.

The inhibition of monoamine enzymatic synthesis is summarized in Table 8.

Table 8[a]

Compound	Inhibits	Action on brain monoamines
α-Methyldopa	5-HTP–dopa decarboxylase[b]	< DA, < NA, < 5-HT
α-Methyl-*m*-tyrosine	L-Tyrosine hydroxylase	< Catecholamines
α-Metatyrosine	L-Tyrosine hydroxylase	< Catecholamines
Disulfiram	Dopamine-β-hydroxylase	< NA
FLA–63	Dopamine-β-hydroxylase	< NA
NSD–1034	5-HTP–dopa decarboxylase	No effects
p-Chlorophenylalanine	Tryptophan-β-hydroxylase	< 5-HT
Salicylhydroxamic acid	5-HTP–dopa decarboxylase	< DA, < NA
Ro 4–4602 (< 50 mg/kg)	5-HTP–dopa decarboxylase[c]	No effects
Ro 4–4602 (> 50 mg/kg)	5-HTP–dopa decarboxylase[b]	< DA, < NA
U-14–624	Dopamine-β-hydroxylase[b]	< NA
MK-485 (DL isomer)	5-HTP–dopa decarboxylase[d]	No effects
MK-486 (L isomer)	5-HTP–dopa decarboxylase[d]	No effects

[a] Represents a summary of a composite of investigations of a series of inhibitors acting on enzymes responsible for monoamine synthesis.
[b] Peripheral and endogenous decarboxylase inhibitor.
[c] Peripheral decarboxylase inhibitor only.
[d] Peripheral decarboxylase inhibitor only, any dose.

3.6. INHIBITION AND RELEASE OF MONOAMINES BY DRUGS

A great deal of information regarding monoamine function in the brain has been derived from psychotropic drugs that affect the cerebral amine storage capacity in one of four ways:

(1) By blocking monoamine incorporation into the storage granules and thus gradually depleting them of the amine transmitter. Compounds such as reserpine, tetrabenazine, and deserpidine may thus mobilize monoamines by causing functional hyperactivity, by impairing the storage sites, by increasing the permeability of particle or cell membranes, or by displacing monoamines from storage sites.

(2) By causing an intraneuronal accumulation of monoamines. Monoamine oxidase inhibitors such as nialamide, isocarboxazid, pheniprazine, pargyline, iproniazid, and tranylcypromine produce this effect.

(3) By blocking the cell membrane pump responsible for amine transport into neuronal cytoplasm. The so-called thymoleptics such as imipramine and amitriptyline may act to concentrate the transmitter in the synaptic cleft, which results in receptor activation.

(4) By blocking monoamine receptors and causing an accumulation

of their 3-*O*-methylated and acid metabolites. Neuroleptics such as chlorpromazine, haloperidol, and butyrophenone tend to diminish the rate of amine passage in and out of the storage pool.

The depleting action of reserpine on 5-HT in brain tissue was first reported by Pletscher *et al.* (1956) and confirmed by Paasonen and Vogt (1956). The mobilizing effect on serotonin by reserpine is not the same in those tissues (enterochromaffin cells, blood platelets, and brain cells) storing the indoleamine. Thus, intravenous injection of 0.1 mg/kg of reserpine in the rabbit causes a marked decrease in brain serotonin levels but not release of the 5-HT store in other tissues to the same extent. The rate of 5-HT depletion is slower in platelets and intestines even when the reserpine dose is increased fiftyfold. Consequently, it takes less than an hour to deplete 50% of serotonin in brain but several hours to deplete the same percentage in extracerebral tissues. Reserpine is also effective in liberating tissues of their catecholamines and appears to manifest an organ specificity similar to its action on 5-HT. This specificity has been demonstrated by Carlsson *et al.* (1957) and Brodie (1957), who observed that about 5 μg/kg single doses of reserpine could release most of the heart catecholamines in rabbits, while 100 μg/kg and 500 μg/kg doses were needed to deplete only 50% of the total catecholamine content in brain and adrenals, respectively. The return to normal values required about 4 weeks for brain amines and 2 weeks for heart and adrenal amines.

Investigation into the pharmacological ability of *Rauwolfia* alkaloids to liberate brain catecholamines and serotonin shows that only those alkaloids capable of causing central depression, as seen in decreased motor activity and alertness, can deplete monoamines from cerebral tissues. Besides reserpine, this effect is produced by other *Rauwolfia* derivatives such as raumescine, rescinnamine, and recanescine but not by isoreserpine, reserpic acid, serpentine, methyl reserpate, reserpinine, or ajmaline.

Many compounds that are not *Rauwolfia* alkaloids exert similar reserpine-like effects on brain monoamines. Such a group is the benzoquinolizine derivatives, as represented by tetrabenazine, a drug that has been shown to act mainly on brain catecholamines and serotonin while affecting only slightly the peripheral amines. The administration of tetrabenazine to various mammals such as rat, dog, cat, rabbit, and monkey produces a depression of spontaneous and induced locomotor activity, as shown by hypothermia, hunchback posture, prolongation of barbiturate anesthesia, enhancement of the convulsant effect of pentetrazol, and, in addition, various autonomic reactions such as miosis, ptosis, lacrimation, and photophobia. The monoamine-depleting effects of the benzoquinolizine derivatives on the central nervous system may be reversed or checked by MAO inhibitors. Still other compounds such as triethyltin, vincamin, α-methylmetatyrosine, α-methyldopa, and metatyrosine can selectively lower one or several brain monoamines. Triethyltin is reported to lower NA and 5-HT levels in the rat brain while not affecting peripherally stored amines except those in the adrenal medulla, where adrenaline is noted to be decreased. The alkaloid structure of vincamin is still unknown, but it is reported to cause reserpine-like effects on the amine levels and behavioral responses in rats.

The effects produced by α-methyldopa and its analogue α-methyl-metatyrosine also mimic the causal activity of reserpine, but monoamines are restored to their normal tissue levels differently. Consequently, the administration of these α-methylated compounds lowers brain dopamine, NA, and 5-HT in animals, but the levels of 5-HT and dopamine quickly rise to normal values as an inverse diminishing inhibition of decarboxylase activity proceeds at the same time. The NA levels in heart and brain in these same animals will remain, curiously enough, markedly decreased for several days. This has led Hess *et al.* (1961) to formulate a "two-mechanisms" theory, in which the initial depletion of monoamines would result from inhibition of the decarboxylating enzyme while the long-lasting effect of NA in tissue might be due to an impairment of the NA binding sites in the tissues. Stone *et al.* (1962) have reported that after the noradrenaline content in peripheral tissues is lowered with α-methyl amino acids, the pressor response to injected amphetamine or phenylalanine is almost abolished in the dog, but that the pressor response to central vagal stimulation is less affected.

This phenomenon is not seen with reserpine, which inhibits both pressor responses when administered in lieu of the α-methyl amino acids. Pletscher *et al.* (1956) and Brodie *et al.* (1966a) have suggested that some of the central actions of reserpine might be mediated through the release of serotonin from its binding sites in brain. Dubnick *et al.* (1960) have argued that the inhibitory action of serotonin in reserpinized animals results only from the alkaloid's lowering of the body temperature and report that if this temperature is returned to normal the reserpine-treated animals will not have a serotonin drop but rather a slight increase in serotonin in brain tissue. This view is also shared by Garattini and Valzelli (1958), who suggest that brain 5-HT levels in the rat increase as the ambient temperature is increased from 22 to 37 C without necessarily changing the animal's body temperature. Carlsson (1964), however, rejects both theories, and instead points out that MAO inhibitors can counteract the sedative effects of reserpine while at the same time increasing the 5-HT levels. If Brodie's hypothesis were correct, Carlsson argues, the reserpine syndrome should be more pronounced due to an increase in available 5-HT, but in reality the sedation effect is abolished. Carlsson proposes that the reserpine symptoms are mainly caused by blockade of the transmission mechanism of the various mono-aminergic neurons. In support of this view, he points to evidence indicating that reserpine produces NA deficiency in brain as well as in peripheral sites and that dopamine overcomes the action of reserpine rather than potentiates its effect in the living organism. Studies show that reserpine causes acute parkinsonian symptoms possibly be due to loss of dopamine from the basal ganglia, which is also seen in patients suffering from "spontaneous" parkinsonism. The syndrome may be relieved by L-dopa, which converts to dopamine in brain and thus may antagonize the reserpine effects.

In addition, histochemical studies indicate that the amine depletion induced by reserpine lasts much longer in the nerve cell terminals than in the cell bodies. The fluorescence seen in the nerve cell perikarya after recovering from reserpine administration is thought to be due to new monoamine storage particles in the Golgi apparatus forming just outside the cell nucleus. The particles and subsequent fluorescence later reappear along the axon, finally spreading to the nerve cell varicosities as the monoamine neurons return to their normal state. As for the action and effects of psychotropic drugs on the central nervous system, Carlsson proposes a hypothetical model of the points of attack and disruption of the drugs on the neurotransmitter system. Carlsson's system contains a so-called membrane pump which is selectively blocked by imipramine-like drugs, especially those containing one methyl group on the nitrogen side chain. Under this condition, drugs which block the membrane pump, as well as MAO cause a loading of exogenous or cytoplasmic NA (or its precursors), which results in extracellular leakage of the amine and subsequent decrease in peripheral and central NA levels. For their part, Brodie et al. (1966a) maintain that reserpine sedation is associated with changes in 5-HT, not catecholamines. Brodie points out that catecholamine synthesis may be blocked and reduced 80% without producing sedation, while reserpine elicits sedation in doses that reduce 5-HT and catecholamines by only 55%. In addition, Brodie finds a time correlation between reserpine effects on behavior and impairment of accumulation of exogenous serotonin in brain tissue.

An interesting observation has been made by Vogt (1965) about two benzoquinoline derivatives with reserpine-like effects previously tested by Weissman and Finger (1962). One of these compounds, an alcohol derivative, depletes catecholamines from the brain, while the other, an acidic acid ester, does not. Vogt raises the question as to whether reserpine treatment in patients really affects monoamines in the brain. However, it is likely that reserpine acts to inhibit the uptake and storage of catecholamines in granules, thus decreasing their concentration in the brain.

The second group of compounds affecting monoamine metabolism in the brain is the monoamine oxidase inhibitors, of which more than 500 have been described to date. The MAO inhibitors are chemically classified into three main types: hydrazine derivatives simple and substituted amines, and certain heterocyclic compounds containing nitrogen.

The first hydrazine compound to be described as a MAO inhibitor was iproniazid, by Zeller and Barsky (1952). This discovery led to the appearance of a great number of related compounds with alkyl and acyl modifications intended to improve organ selectivity. This selectivity may be seen by comparing iproniazid and β-phenylisopropylhydrazine (PIH, JB-516) in the normal rat. These agents subcutaneously administered show an equipotent inhibitory activity against liver MAO, but in brain PIH inhibits MAO at much lower doses. This preferential action of PIH on the brain enzyme has encouraged further work on other alkyl- and arylalkylhydrazines which could result in further brain selectivity. At the brain level, however, the problem is complicated by the fact that organ or tissue specificity is not only dependent on the structure of the inhibitor and its lipid solubility and subsequent penetration through the blood–brain barrier but also on the onset of the drug's action and the route of administration. This last phenomenon has been studied (Horita and McGrath, 1960) by administering subcutaneous doses of PIH and phenelzine to rats, which results in a greater MAO inhibition at the brain level than in the liver. The selectivity is reversed when the drugs are orally administered. A third response is obtained when the two drugs are given intraperitoneally, the brain and liver enzymes being equally inhibited. In contrast, compounds such as nialamide and iproniazid exert a greater MAO inhibition in the liver regardless of the route of administration. Time dependence of MAO inhibition may be produced by the increased levels of monoamines in the various tissues. The time interval required for the monoamines to augment their levels is therefore partly an aspect of the drug's function and its dosage concentration.

With respect to the ability of MAO inhibitors to penetrate the blood–brain barrier, Bertler et al. (1966), using biochemical and histochemical methods, have demonstrated the existence of a specialized brain barrier mechanism at the vascular level that permits or impedes the passage of L-dopa into the brain parenchyma. Intracerebral vessels thus appear to contain the enzyme MAO and dopa decarboxylase, which will attack or block administered MA precursors unless the animal is previously treated with MAO inhibitors or decarboxylase inhibitors. Blockade of MAO by inhibitors leads to the reversal of tetrabenazine-induced central depression and a decrease in monoamine levels. Moreover, the reserpine syndrome is antagonized in animals and man regardless of whether the inhibitor is given before, during, or after reserpine. The clinical application of treatment with MAO inhibitors lies in their antidepressive action, which inhibits the degradation of monoamines and consequently produces an increase of NA and 5-HT in the brain and various peripheral organs. In various animal species studied (Table 9), the MAO inhibition and resulting catecholamine and serotonin increase are accompanied by behavioral excitation, which may be linked to the impairment of normal metabolic deamination that is

**Table 9. Dose of MAO Inhibitors (mg/kg)
Injected Intraperitoneally into Various
Species of Animals and Producing a
50 % Increase in Brain Serotonin[a]**

	Rabbit	Guinea pig	Mouse	Rat
Iproniazid	290	64	540	175
Isocarboxazid	74	14	44	28
Pargyline	—	224	405	81
Tranylcypromine	—	28	23	6

[a] Killed after 16 hr.

produced by the inhibition of MAO, which normally inactivates exogenous monoamines. In addition, Spector *et al.* (1960) have demonstrated disparate effects from using different inhibitors and varying their doses in animals. Elevation of NA and 5-HT after treatment with MAO inhibitors was correlated with behavioral excitation, but in animals in which 5-HT was increased without affecting NA levels, no behavioral excitation occurred. This, as well as the reserpine antagonism by MAO inhibitors, which appears to be better correlated with NA than 5-HT changes, is dependent on free NA leaking out of its storage site to activate adrenergic receptors. Excitation results when the MAO enzyme is subsequently inhibited by a proper agent and hence cannot attack or inactivate the unbound NA responsible for the adrenergic stimulation. Reserpine treatment, however, does not affect MAO, and when loss of NA from its storage site occurs, the amine is rapidly inactivated by MAO and sedation results.

It must be recalled that this theory has not yet been proved, due in part to the technical difficulty of isolating the receptor system, which could allow a direct analysis of the amine–receptor interaction. Work by Matsuoka *et al.* (1965) using 1-(5,6-dimethoxy-2-methyl-3-indole)-ethyl-4-phenylpiperazine (WIN 18501-02) appears to support the above view. WIN 18501-02 reduces the NA levels *in vivo* without affecting serotonin, and 3 hr after injection of the drug there is a decrease of granulated vesicles in the anterior hypothalamus, which is thought to be the principal storage site for NA. Sedation and loss of these granulated vesicles are well correlated both after WIN 18501-02 and after reserpine. Spector (1963) has indicated a species difference in behavioral reaction to the nonhydrazide MAO inhibitor pargyline (MO-911). After administration of MO-911 to cats and dogs, no excitation or rise in brain NA levels was noted, but relative treatment in rabbits resulted in a positive reaction, that is, an elevation of NA in brain and central behavioral excitation. In human clinical trials, Maclean *et al.* (1965) recorded a serotonin increase with nialamide or iproniazid administration only after 2 weeks of treatment, at which time the 5-HT levels rose to twice their normal values and remained at that concentration even after the two MAO inhibitors were continued for another 2 weeks. Nialamide and iproniazid also elevate the adrenaline content in the rat brain and heart, but nialamide also increases the NA concentration in these two organs. As for serotonin, Weber (1966) contends that in rabbit brain iproniazid produces an increase, but not as marked as when nonhydrazine MAO inhibitors such as MO-911 or N-methyl-N-2-propynyl-benzylamine HCl are given in relative doses. Further work using iproniazid indicates that the drug causes a slow but apparent rise in brain catecholamines, a specific increase in hypothalamic NA, and a stronger dopamine rise in the neostriatum. The data suggest that iproniazid treatment has an inhibitory effect on brain and liver MAO with a subsequent increase in the concentration of monoamines.

The MAO inhibitors have certain pharmacological characteristics which may be summarized as follows: (1) They are mood elevators or antidepressants in man. (2) They potentiate the pharmacological effects of monoamines, perhaps by causing their intraneuronal accumulation and increased concentrations in brain. (3) They antagonize in many cases reserpine-like effects. (4) They may be used as hypotensive agents. (5) In addition, after prolonged treatment with MAO inhibitors in humans, some central effects have been observed. These include sleeplessness and agitation. Exacerbation of symptoms has been recorded in schizophrenics given MAO inhibitors in long-term therapy as well as delirious and confusional states. Extrapyramidal motor functions may be impaired and result in hyperkinetic chorea. Various suicides have been reported using MAO inhibitors; Hollister (1964) mentions a 17-year-old girl who committed suicide by ingesting 500 mg of tranylcypromine. Delirium, tremors, coma, and shock preceded heart block and hyperthermia. Death occurred within 8 hr after taking the drug. The exact relationship, if any, between the MAO inhibitors and monoamines in cerebral function and central effects remains to be explained.

A third group of psychoactive compounds, the thymoleptics, includes imipramine and other tricyclic derivatives. Clinically, these drugs have been found to be the most effective antidepressant compounds, although in the case of imipramine there is no inhibition of MAO or catechol-O-methyltransferase. Glowinski and Axelrod (1964) have found evidence that imipramine and two other tricyclic analogues, desmethylimipramine (DMI) and amitriptyline, inhibit NA uptake in the brain. In peripheral tissues, infused NA in imipramine-treated animals is not taken up normally. This suggests the possibility that imipramine blocks or decreases the cell membrane or storage granule membrane permeability to this catecholamine. The antidepressant action of imipramine may be due in part to inactivated free NA sensitizing the central adrenergic synapses. This potentiation of NA could lead to a decrease in the inactivation rate of the free amines due to their storage under the influence of the thymoleptics. In contrast, the MAO inhibitors are capable of potentiating various adrenergic functions through the inhibition of catecholamine degradation. Aside from this action, the thymoleptics can also reverse the sedative effect of reserpine and some other benzoquinolizine derivatives. If the MAO inhibitors inhibit the symptomatic and biochemical effects of reserpine, the antidepressants inhibit only the symptomatic effects, without changing the MAO concentration in the brain (see Table 10). Sulser et al. (1964) have also confirmed the antagonistic reaction of DMI toward the benzo-quinolizines and reserpine, affirming the hypothesis of thymoleptic potentiation of the amines. This antagonism may be due to the rate of brain noradrenaline release, as indicated by the finding that when animals are partially depleted of their NA stores by α-methylmetatyrosine, pretreatment with DMI does not antagonize reserpine or tetrabenazine sedation. The antidepressants, moreover, possess a variety of anticholinergic properties. The anticholinergic actions of the thymoleptics are seen both in vitro and in vivo, producing mydriasis and hyposecretion of saliva. The anticataleptic effect of these drugs is further evidence of central anticholinergic activity.

Schanberg et al. (1967) found that noradrenaline-H^3 administered intracisternally into rat brain increased brain concentrations of the O-methylated metabolite normetadrenaline-H^3 when the animals had been pretreated with imipramine or DMI but not with chlorpromazine (CPZ), a similar-acting tranquilizer. In another study, Eisenfeld et al. (1967) showed that drugs which block adrenergic receptors in peripheral tissues decrease the formation of normetadrenaline.

Table 10.   Effect of Intraperitoneal Imipramine and Amitriptyline on Endogenous 5-HT and NA in the Brain of Rats [a]

	Controls	20 mg/kg		50 mg/kg	
		1 hr[b]	4 hr[b]	1 hr[b]	4 hr[b]
		Imipramine			
5-HT	0.60 ± 0.02	0.63 ± 0.03	0.60 ± 0.03	0.66 ± 0.02	0.60 ± 0.02
NE	0.26 ± 0.04	0.26 ± 0.04	0.24 ± 0.02	0.24 ± 0.04	0.26 ± 0.02
		Amitriptyline			
5-HT	0.60 ± 0.02	0.68 ± 0.05	0.55 ± 0.02	0.65 ± 0.03	0.63 ± 0.05
NE	0.26 ± 0.04	0.26 ± 0.04	0.30 ± 0.02	0.22 ± 0.02	0.26 ± 0.02

[a] The figures indicate the 5-HT and noradrenaline (NA) content in μg/g fresh brain; mean values with standard error. No significant changes in the amine concentration after drug treatment are evident. From Pletscher and Gey (1962b).
[b] Hours after injection of drugs.

Imipramine is also capable of diminishing in rabbits the severity of seizures produced by electroshock or by pentetrazol, but the same convulsions are not affected by CPZ. Finally, there has been demonstrated the inhibition of uptake of radioactive NA in the rat brain by DMI, imipramine, and amitriptyline, but not by CPZ. Thus, at least two mechanisms of action of the thymoleptics on NA may exist: diminishing or blocking of the inactivation process at adrenergic receptors and inhibition of NA uptake into storage granules.

From most of the available data on the thymoleptics, three properties of these drugs are evident: (1) sedative–tranquilizing effects and slight adrenolytic action; (2) potentiation of catecholamines and inhibition of NA rebinding, resulting in an increase of free NA at the level of the receptor organs; and (3) an anticholinergic effect on central and peripheral tissues.

The last group of compounds is the neuroleptics or major tranquilizers; of the 50 or so drugs used in clinical practice, chlorpromazine and haloperidol are the best known. The neuroleptics induce a wide variety of actions similar to those of reserpine but do not change amine levels or the storage mechanism, mimicking in this sense the action of antidepressants. One of the striking characteristics of neuroleptics is their ability to counteract hyperexcitability brought on by the administration of monoamine precursors. Following treatment with monoamine precursors, there appear certain psychomotor symptoms such as aggressiveness, characteristic head movements, and other effects apparently involving extrapyramidal disturbances. The capacity of neuroleptics to arrest this syndrome is, according to some workers, further evidence that the drugs block the receptor system in the central nervous system.

According to Pletscher and Da Prada (1967), the neuroleptics can be classified into two types: those whose action is independent of hypothermia and those that function only with a rise in body temperature. These same two workers have shown that neuroleptics block the uptake of intravenously administered NA in the hypothalamus and certain other tissues, increasing at the same time NA metabolites (e.g., normetadrenaline) in the blood. Pletscher and Da Prada also indicate that CPZ derivatives increase the content of the dopamine metabolite homovanillic acid (HVA) in the brain without altering the endogenous dopamine content in that organ. This mechanism might be described to be the result of dopaminergic interference at the receptor sites by the neuroleptics leading to a compensatory increase in dopamine synthesis and a subsequent rise in HVA outflow from the brain. The increase in HVA outflow is not due to hypothermia. Johnson (1964) took normal (30 C) and cold-acclimatized (2 C) rats and compared them with a second group exposed to the same temperatures but pretreated with CPZ. In the nontreated group, no increase in catecholamine excretion was noted, while the second group of animals showed an increase in NA and a lesser increase in adrenaline excretion at both 30 and 2 C. This increase in catecholamine excretion has been suggested by Johnson to result from a temporary hypothermia induced by CPZ and originating from sympathetic nerve endings. Hypothermia associated with adrenalectomy also leads to increase in catecholamine excretion but may be prevented by pretreatment with a ganglionic blocker. The ability of reserpine to lower serotonin and noradrenaline values in brain is inhibited by treatment with CPZ, which also inhibits an increase in the two monoamines after treatment with MAO inhibitors. If the serotonin precursor 5-HTP is given with CPZ, there is likewise an inhibition of brain serotonin increase. Since CPZ does not appear to inhibit the action of MAO or the dopa decarboxylase enzyme, Pletscher and Gey (1962b) have suggested that CPZ and similar derivatives may function by decreasing the permeability of monoamine storage granules. This theory has found support in the work of Nathan and Friedman (1962), who observed a decrease in the permeability of resting cells when CPZ was administered to tetrahymena pyriformis. Histological studies comparing several neuroleptics with the tranquilizers chlordiazepoxide (Librium) and diazepam (Valium) indicate that the latter drugs produce mainly cytocortical changes. The action of the neuroleptic haloperidol is more widespread, producing changes in the thalamus and mesencephalon, while CPZ induces mild and reversible cellular damage, frequently involving the nucleus (see Table 11). From a study comparing neuroleptics and tranquilizing drugs, Pletscher and Da Prada (1967) concluded that while the two groups have quite different effects on the biochemistry of brain catecholamines, there does not appear to be any difference in their metabolism.

The interesting experiments by Olds *et al.* (1957) on self-stimulation of the brain by animals with chronically implanted electrodes show the specific action of CPZ on brain regions: (1) major effect on the posterior hypothalamus with electrodes deep in the medial forebrain bundle, (2) little effect on the dorsal and anterior hypothalamus, and (3) strong effect on the posterior parts of the forebrain.

Stein (1968) has advanced the theory that in the mammalian brain reward and punishment mechanisms are mediated by two pathways, the medial forebrain bundle for reward and a periventricular system of fibers for punishment. It is conceivable, if one accepts the above premise, that drugs with specific action on one of the two pathways could induce or inhibit pleasant or unpleasant neural events. This may lead in the future to the development of the science of "neuroeupharmacology."

3.7. SLEEP MECHANISMS

It is estimated that 33% of man's life is spent in sleeping. The process and nature of sleep have puzzled and fascinated man since early times.

Table 11. Cytological Changes Induced by Neuroleptics and Tranquilizing Drugs [a]

Drug	Dose[b] (mg/kg)	Cortex	Brain stem
Chlorpromazine (CPZ)	25	(a) Pale cytoplasm (b) Nuclear membrane tortuous (c) Scattered chromatin	(a) Dimorphism of motoneurons (b) Nissl substance fragmented
Haloperidol	0.8	Thalamic dysmorphism of nuclear membrane; hyperchromia	
Chlordiazepoxide (CD)	10	(a) Swelling and paleness of axons (b) Increased cytoplasmic volume and pyramidal cells	(a) Slight chromatolysis in nucleus of motor neurons
Diazepam	2.0	(a) Swelling of cytoplasm and axons	(a) Slight hypochromia

[a] Data from Cazzullo (1967).
[b] Doses given daily for 1 month. CPZ and haloperidol administered intraperitoneally; CD and diazepam administered subcutaneously.

Nevertheless, the scientific exploration of sleep phenomena by electrophysiological, biochemical, and neurological approaches has been a relatively recent development, but one that has lately yielded a number of new concepts and findings. Today, with tools such as the electroencephalograph, electro-oculograph, and electromyograph, various mechanical and motor functions in man and animal may be tested during the different phases of sleep activity. Biochemical testing in man, however, is rather limited, as only urine or blood can be analyzed for metabolic changes. In animals, study of tissue samples from the central nervous system has resulted in new insights into the biochemically activated changes during sleep.

Many years before the advent of the neurophysiological and biochemical techniques presently used in sleep investigation, dream interpretation as expounded by Freud (1955) had become an important psychotherapeutic tool in clinical psychiatry. His theory was that dreams as interpreted by the analyst permit discovery of the basis of behavioral problems. Events occurring in dreams were symbolized by the subject, and it remained for the therapist to unravel their hidden meanings and then proceed to administer an appropriate form of therapy. Dreams are still an important factor in psychoanalysis; however, symbolic dream interpretation has so many serious shortcomings that so far as its application to improving our understanding of dream mechanics goes, it has become as irrelevant as the Greenback party in the United States. If the past is prologue, then it is evident that bad theories will not fall by criticism or by their lack of validity in controlled studies. Rather, they will be replaced by better theories. Dream interpretation is based on three improbable or, at best, unproven premises: (1) that inductive analysis of events dreamed is feasible in subjects of different cultures, motivations, and values; (2) that the subject accurately relates the dream context to the analyst; and (3) that the subject "symbolizes" his thoughts in dreaming. The complexity of human thought and emotion is sometimes equally reduced to absurdity when efforts are made to relate rapid eye movements in sleeping animals to a "dreaming" stage. Dreaming is known to occur only for *certain* in man.

It is not the purpose of this book to report on the vast literature on this similar subjects but merely to discuss the basis of the present knowledge about sleep mechanisms and how they relate to the monoamines. The idea that monoamines may be implicated in sleep resulted from a series of experiments done by Jouvet and some of his doctoral candidates in Lyon, France, about 5 years ago. Since then, a number of publications have confirmed the findings that noradrenaline is related to paradoxical sleep (PS) or rapid eye movement, and that 5-HT is active in slow sleep (SS) activity. The classic experiments of Aserinsky and Kleitman in 1953 and Detoni in the same year showed that during sleep in humans, eye movements appear to be related to the depth of sleep. Aserinsky and Kleitman used the technique of continuous observation of their subjects, and with the electrooculogram they were able to follow the sleeping cycle which involved rapid eye movements. If the subjects were awakened during the rapid eye movement phase, 78% of them reported an interrupted dream. When they were awakened at any other time during their sleep, no recollection of a dream was reported.

The electroencephalogram (EEG) was used by Dement and Kleitman (1957) to try to correlate cortical wave activity with the rapid eye movement phase. This relationship had been previously reported by Klaue in 1937. Klaue showed that slow sleep and slow cortical activity followed by PS with fast cortical activity were related. EEG recordings during complete sleep cycles merely confirmed other findings that during certain stages of this cycle, there occurred a phase of fast cortical activity which was immediately related to dreaming.

The Swiss Nobel Laureate Walter Hess (1965) advanced the notion that sleep is a parasympathetic function of behavior. According to his classification of cerebral mechanisms actuating autonomic responses, the "trophotropic" component (consisting of subjective signals in living organisms such as hunger and thirst) would mediate sleep activity. This is interesting in the light of Brodie's extension of Hess' "tropic" zones in which he postulated that catecholamines may mediate the sympathetic or ergotropic zone while serotonin may be responsible for parasympathetic or trophotropic responses.

There are at the present time three states of consciousness known in man, waking, sleeping, and dreaming (Fig. 27). The evidence for a monoamine role in the latter two states will now be examined.

3.7.1. Slow Sleep and Serotonin

Slow sleep is characterized in EEG recordings by the appearance of slow waves and spindles at the cortical and subcortical levels. Somatic signs are nuchal muscle activity without recordable rapid eye movements. About 70% of total sleep is taken up by the SS phase.

Most of the evidence linking serotonin with SS comes from the work of Jouvet and his colleagues. Their findings suggest that when brain 5-HT

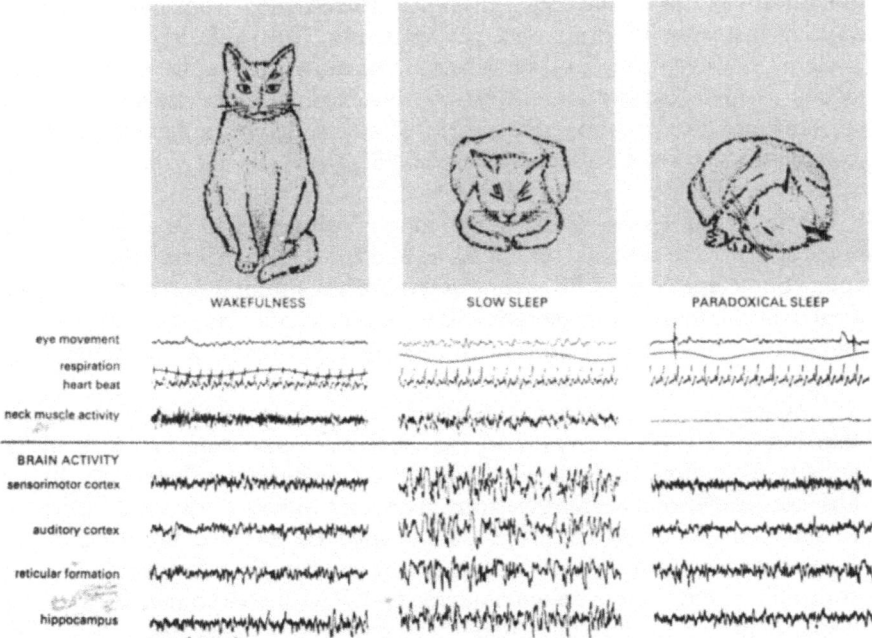

Fig. 27. Three states of the nervous system. During wakefulness, there is fast, low-voltage cortical activity. During slow sleep, the activity is slow but much more pronounced, giving rise to higher-voltage EEG (bottom four traces). During paradoxical sleep, EEG patterns are the same as for wakefulness but there are rapid eye movements and a total disappearance of neck muscle tone. From Jouvet (1967c).

is increased, SS time is also increased. Inversely, when 5-HT is decreased, SS is diminished or abolished. Moreover, it would appear that cerebral 5-HT levels are proportional to the amount of SS time expended. To support this relationship, Jouvet (1968) points to the raphé complex as the most dense concentration of 5-HT-containing neurons. These cells extend from the nucleus linearis in the mesencephalon to the raphé obscurus in the medulla. The cells belonging to this system have been described by Brodal (1970) as forming nine groups (B1 to B9). Histochemical fluorescence has confirmed the existence of these 5-HT neurons in the regions described (see Section 3.2). When the cells are destroyed, terminal fluorescence allegedly ascending rostrally from these 5-HT cells disappears after a week. The same is true for spinal cord terminals efferent to the raphé cells. In animals, when selective stereotaxic lesions were made in some of the raphé groups, a constant EEG recording revealed a loss in SS time proportional to the amount of cells destroyed. Subtotal destruction of the raphé system (80–90 %) by monopolar coagulation brought on virtual insomnia in the animals tested. The animals seemed to be always in motion, and they had tachycardia and pupillary dilatation. No slow wave or spindles were recorded from cortical EEG; indeed, fast wave rhythms were predominant. Total ablation of the raphé system, however, can not be effected, because the raphé pallidus and obscurus are linked to vital respiratory centers. Sham-operated animals, that is, subjected to stereotaxic placement of electrodes in raphé cells without coagulation, did not suffer any loss of SS. These animals had normal SS activity, averaging 12 out of 24 hr.

With partial lesions of the raphé groups, either a half or a third of the nuclear area, the animals remained awake for 2 days. SS reappeared soon after. Some episodes of SS were observed after subtotal destruction of the raphé on the fourth postoperative day, but these only lasted for about 1 hr out of the whole day. At no time did SS last more than 3 hr during the day in animals with subtotal lesions.

When 5-HT was measured after partial or subtotal surgery, it was found that loss of 5-HT was related to the amount of insomnia and lack of SS activity. Variations in the level of noradrenaline did not influence SS. In the absence of neuroanatomical lesions, a second line of evidence relating SS to serotonin comes from drugs that change the brain concentrations of serotonin selectively. For example, serotonin can be reduced by 90% of control levels by p-chlorophenylalanine (p-CPA), a compound that inhibits tryptophan hydroxylase and therefore the synthesis of serotonin centrally. Doses of 100–300 mg/kg of p-CPA given to cats (Delorme et al., 1966) are reinforced after 24 hr by a second injection. Total insomnia follows on the third day after the first injection of p-CPA. The animals remain awake for 2 days without any visible excitation. Normal SS activity returns after 10 days following p-CPA. A second drug, p-chloromethamphetamine, at

10–20 mg/kg in cats, produces insomnia 1 hr after administration. The animals are generally agitated compared with those treated with p-CPA. The return to normal sleep pattern is seen after about 60 hr. However, Wyatt *et al.* (1969) examined the effects of p-CPA (3–4 g/day) in four human subjects and found no changes in SS activity after 10 days of treatment.

Increased SS time, according to Jouvet (1968), can be achieved chemically by the administration of 5-hydroxytryptophan (5-HTP) or nialamide. The former is, of course, the direct precursor of serotonin, while the latter is a MAO inhibitor that increases brain 5-HT. Jouvet reports that 40–50 mg/kg intravenous doses of 5-HTP suppress PS and its characteristic ponto-geniculo-occipital (PGO) spikes for 3 hr. Matsumoto and Jouvet (1964) had previously shown that when reserpine, 0.5 mg/kg, was given to to cats, a loss of SS resulted, lasting for 2 days. However, injection of 5-HTP, 50 mg/kg intravenously, in previously reserpinized animals abolished PGO spikes, and SS reappeared uninhibited. It is hard to comprehend Jouvet's data linking administered 5-HTP to increased SS activity. His contention is that if 5-HTP is injected into animals, a rise in *endogenous* brain serotonin will result, and he concludes (Jouvet, 1968) that a rise in the endogenous brain 5-HT will directly affect the duration of the SS phase, in this case lengthening it. It has, however, been shown in several recent studies (de la Torre, 1968b; Constantinidis *et al.*, 1968; Bertler *et al.* 1966) that the blood–brain barrier penetration of the monoamine precursors 5-HTP and dopa is dependent on the peripheral dopa decarboxylase activity in cerebral capillaries. If peripheral dopa decarboxylase is not inhibited prior to treatment with the 5-HT precursor, the administered 5-HTP will be quickly decarboxylated to serotonin at the capillary level. The serotonin in the brain capillaries will then be gradually deaminated by MAO, which is also present in the capillaries. It is obvious that if biochemical assays are done on homogenized brain tissue after 5-HTP injections, a rise in brain serotonin *will* be obtained but only as a reflection of *peripheral* serotonin, since the capillaries containing the newly formed amine will be homogenized nonselectively along with the brain tissue. The peripheral accumulation of 5-HT and subsequent metabolic changes also affect the urinary excretion of 5-hydroxyindoleacetic acid, so trying to determine 5-HT increases in brain by measuring this is also useless. These simple errors have led to the belief that monoamine precursors do in fact cross the brain barrier without difficulty, and it is often quoted in the recent literature.

The next part of Jouvet's experiments centered around increasing brain 5-HT by prior administration of a MAO inhibitor. Nialamide, a strong MAO inhibitor, was used for this purpose. Jouvet (1967c) reports that after 10 mg/kg animals show high-amplitude cortical spikes which last 5–6 hr. Following this, cortical spikes are seen with high-voltage, slow wave activity. SS then suppresses PS and occupies 60–80% of the total sleeping time. This phenomenon, according to Jouvet, lasts for 3 days, until PGO waves reappear for 1–2 min. On day 4 after the initial dose of MAO inhibitor, the sleep pattern returns to normal.

In his final argument, Jouvet proposes that any selective decrease in brain 5-HT, whether by mechanical or pharmacological means, results in a suppression of SS, while the reverse (i.e., an increase in 5-HT) leads to an increase in SS. Furthermore, no compensation of PS is found in animals after treatment with MAO inhibitor. Using MAO inhibitor on humans, Wyatt et al. (1969) found some suppression of PS after relatively high doses. However, they found no changes in SS as Jouvet did in cats. Moreover, there appeared to be PS compensation in subjects studied for 10 or more days. This compensation of increased PS after its suppression has also been shown by Dement (1966) in human volunteers and ranges from mild to strong compensatory increases.

3.7.2. Paradoxical Sleep and Noradrenaline

We are indebted to Jouvet for coining the term *paradoxical sleep* (PS), which replaces other more vague descriptions such as *rapid eye movements, D-state, activated sleep, LVF sleep, hindbrain sleep,* and so on *ad nauseam.* PS is characterized by fast, low-voltage cortical activity in the form of ponto-geniculo-occipital (PGO) spikes. Electromyogram activity in the neck is abolished (muscular atony) during PS, and emerging rapid eye movements become the principal phasic sign.

Jouvet's experiments have opened possible new lines of research into the relationship between noradrenergic neurons located in the locus coeruleus and PS duration. Bilateral destruction of the locus coeruleus in the cat is said to suppress PS without marked changes in SS activity for at least 2 weeks (Roussel, 1967). When the noradrenaline precursor dihydroxy-phenylalanine (dopa) is given to animals at a dose of 50 mg/kg, a state of wakefulness occurs for 5 hr. The PGO spikes follow normally after this interval. Dopa also has the capacity, according to Peyrethon-Dusan (1968), to reverse the effects of reserpine and to increase the interval of PGO spikes. In the cat, reserpine appears to inhibit the tonic signs of PS and at the same time to trigger the PGO spike responses which are characteristic PS (Jouvet, 1968). In man, reserpine has the opposite effect. In doses of 0.02 mg/kg, reserpine increases PS time and probably (Hartmann, 1968) decreases total SS time. Another drug, α-methylmetatyrosine, displaces noradrenaline from its storage site, resulting in a reduction of this catecholamine. In doses of 100–200 mg/kg, α-methylmetatyrosine, which is also said to act as a "false transmitter," can inhibit PS in the cat for 12 hr, at the same time increasing SS. This biphasic effect can also be reproduced with some α-adrenergic blockers such as dibenamine and phenoxybenzamine. In contrast to the PS suppression by locus coeruleus lesions, destruction of the pontine tegmentum or the area rostral and caudal to the locus coeruleus does not (Mouret and Delorme, 1967) cause any changes in the pattern of PS activity in rats. In cats, however, such lesions of the pontine tegmentum are reported (Jouvet, 1968) to abolish PS selectively.

MAO inhibitors, discussed in Section 3.5.1, appear to suppress PS and enhance SS in cats. After 10 mg/kg of nialamide, a potent MAO inhibitor, cats no longer show PGO phasic activity. Pupil dilatation and inertia accompany the phasic signs (Jouvet, 1968). SS is enhanced, according to Jouvet, due to increased endogenous 5-HT, which in turn suppresses PS activity. This logic seems contradictory. If PS time is proportional to noradrenaline levels in the pontine region, then drugs which increase noradrenaline would also increase PS. This does not appear to follow with the MAO inhibitors, which increase 5-HT as well as NA in the brain.

Another drug that has selective action on the synthesis of noradrenaline is disulfiram. One of the actions of disulfiram is inhibition of dopamine-β-hydroxylase, the enzyme responsible for the transformation of dopamine to NA. Peyrethon-Dusan (1968) described the use of disulfiram (400 mg/kg, orally) in cats. Hypotonia, anorexia, pupillary dilatation, and neurological signs such as tremors, athetosis, and ataxic gait were observed. Electro-encephalographic recordings showed a slight decrease of SS from the normal values, while marked changes in PS lasting 24 hr were seen. Disulfiram appears to affect especially the rate and amplitude of PGO spikes. Normal PGO spikes, which occur 24 hr following disulfiram therapy, can be restored within 3 hr with reserpine.

A number of other drugs have been used which affect either PS or SS or both, and these are summarized in Table 12.

The evidence to date strongly indicates a monoamine role in sleep activity. What this role is and how it is performed should be topics of future research in this area. As yet to be determined are such issues as the exact function of PS. Experiments on the tortoise show that the EEG is totally lacking in PS activity. In chickens, pigeons, and hens, PS rhythm is record-able for only a few seconds. In hibernating animals, only slow wave activity barely recordable at the cortical level is observed. How brain monoamines are affected during hibernation in bats was the object of a recent study by Constantinidis et al. (1970). Hibernation is a hypothermic state seen in some mammals, resulting in a reduction of energy expenditure and a depres-sion of higher nervous center activities. In bats, as in other poikilothermic hibernators, hunger and cold are the basic factors that trigger this state of "winter sleep." In these animals, hibernation is a highly developed mechanism that allows them a temporary reprieve from food-seeking activ-ity and metabolic energy consumption. The metabolic rate among bats is so high that without food they cannot maintain their normal body temperature

Table 12.   Effect of Some Drugs on Paradoxical Sleep (PS) and
            Slow Wave Sleep (SS) Activities in Subprimates[a]

Compound	PS	SS	mg/kg	Animal
Physostigmine	●	□	Variable	Cat
Atropine sulfate	△	□	0.5–2	Dog, rabbit, cat, rat
Acetylcholine	●	●	Variable	Cat
Carbachol	●	□	Variable	Cat
Pilocarpine	●	○	Variable	Cat
Nicotine	□	○	Variable	Cat
LSD-25	○	□	0.002–0.005	Cat, rabbit
LSD-25	●	□	0.001	Rat
Imipramine	△	●	2–4	Cat, rat
Chlorpromazine	△	□	4	Cat
Amphetamine	△	△	2–4	Cat
Phenylisopropylhydrazine	△	●	1–4	Cat
p-Chlorophenylalanine	△	△	200–400	Cat
Nethalide (β-adrenergic blocker)	□	□	Variable	Cat

[a] In man, the effects of some of the above drugs have been recorded; they do not
   always mimic the animal responses, indicating some significant species variability.
[b] Key: no action □, suppress △, diminish ○, and increase ●.

for any length of time. Inasmuch as inactivity in these mammals results
from a rapid drop in body temperature and metabolic rate, hibernation
appears to constitute an important factor in their survival.

Normal and hibernating bats were killed, and brain monoamines were
examined using the histochemical fluorescence technique. The findings
can be summarized as follows:

(1) The periarterial adrenergic innervation showed a marked decrease
in green fluorescence (noradrenaline) in the hibernators as compared to the
controls (Fig. 28).

(2) There were no changes in the striatal fluorescence of hibernators and
controls.

(3) A stronger catecholamine fluorescence was noted in the tuberal
region of the hibernators. In the controls, catecholamine-containing cell
bodies and synaptic terminals were normally scattered in various hypothalamic nuclei. The catecholamine terminal fluorescence in the hibernators was localized as an intense lamina surrounding the region of the
third ventricle. In addition, no cell body fluorescence was seen in the
hibernating bats anywhere in the hypothalamus (Fig. 29).

(4) The catecholamine cell bodies were considerably reduced in number
in the hibernators in such regions as the locus coeruleus, locus niger, and
lateral nuclei of the pontomedullary region (nucleus paraolivaris and
nucleus reticularis lateralis).

(5) In the pontomedullary raphé system, numerous and intense yellow fluorescent cells (serotonin) were noted in the hibernators as well as the absence of catecholamine nerve terminals. By contrast, control bats showed few, weakly fluorescent serotonin-containing cells but abundant catecholamine terminals in those same regions.

If hibernation is considered a type of sleep but of long duration, these results would appear to agree with Jouvet's data on SS and serotonin content in the raphé. The raphé system may work continuously during hibernation, and the neurochemical expression of that work could be the serotonin concentration. The overworked system could be an exaggerated reflection of the normal sleep mechanism. In normal sleep, however, changes in the serotonin content may be so small that detection could only be achieved by sensitive quantitative assays. The exaggeration of this phenomenon during hibernation would explain why visualization of serotonin changes is possible.

The catecholamine reduction in the periarterial adrenergic innervation appears to support Draskóczy and Lyman's (1967) findings of greatly greatly decreased sympathetic activity in hibernating animals. The periarterial reduction could also be explained by the lack of activity during hibernation and the unapparent need of the animal to control cerebral blood flow. The periventricular laminar concentration of catecholamine terminals and increased catecholamine tuberal fluorescence could be related to the hypothalamic control of body temperature as the animal enters

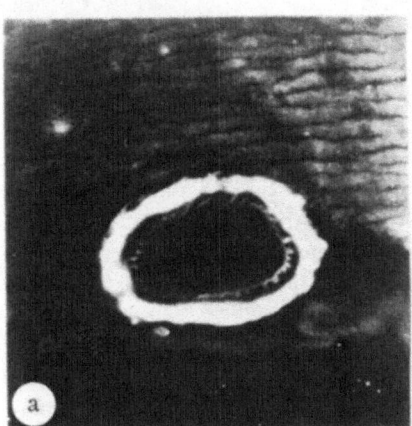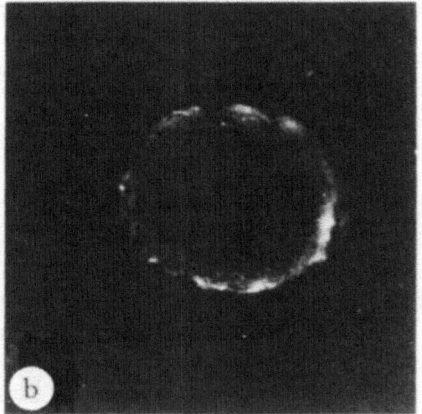

Fig. 28. Periarterial adrenergic fluorescence in normal awake (a) and hibernating bats (b). Decrease of periarterial fluorescence in (b) is accompanied by marked decrease of catecholamine fluorescence in the lateral medullary nuclei and locus coeruleus neurons as well as an increase in yellow (serotonin) fluorescent cells in the mesencephalic raphé system as compared to awake controls. See text for explanation. 100 × . (Reduced for reproduction 10%.)

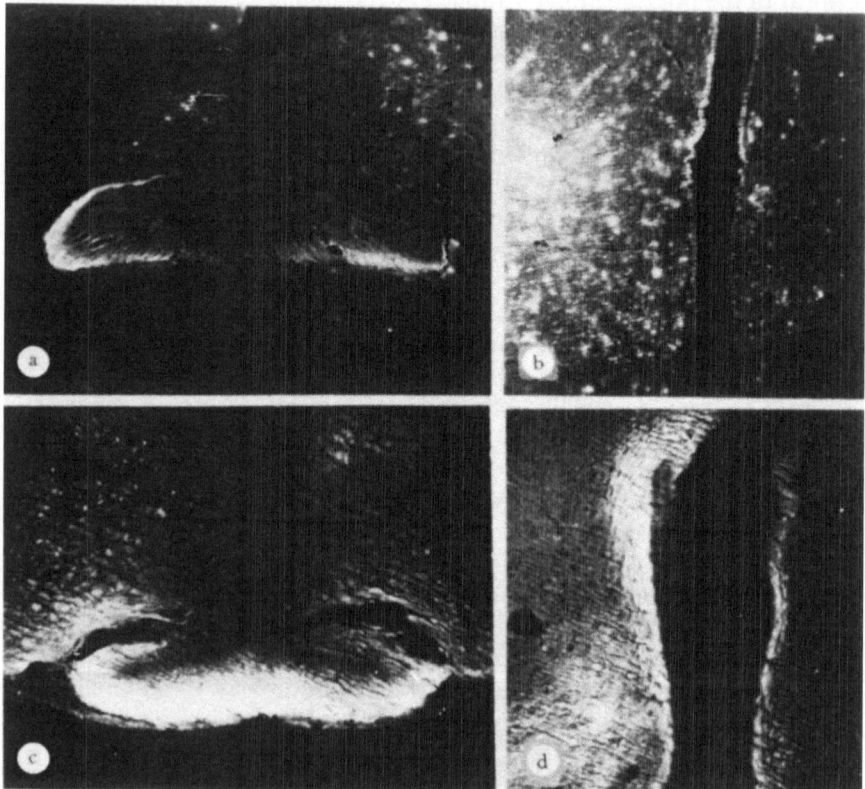

Fig. 29. Catecholamine fluorescence in normal awake and hibernating bats. Decrease in tuberoinfundibular fluorescence (a) and granular periventricular fluorescence (b) in hypothalamus of hibernator. Normal awake bat shows typical strong tuberoinfundibular fluorescence (c) and diffuse periventricular "laminar" fluorescence (d). See also Fig. 29.

hibernation. Finally, the reduction in the number of cells at the locus coeruleus and lateral nuclei in the hibernators may parallel the phenomenon seen during PS; Kayser (1961) and Satinoff (1970) reported an EEG activity characteristic of PS in hibernating animals.

Other evidence which favors a monoamine role in sleep mechanisms has been reported by Brooks (1968). PGO spikes were monitored in the oculomotor and visual systems of awake cats or during rapid eye movement sleep. This phasic activity was not seen during slow wave sleep. After treatment with the 5-HT depleter p-CPA, the fast wave activity seen in the awake state became very great, a phenomenon that could be reversed with large doses of 5-HTP. Methysergide, which antagonizes brain 5-HT, has been reported by many clinicians to produce insomnia in patients treated with this drug for migraine headache. In hamsters, increased motor activity during the awake state is said to decrease brain 5-HT, while 5-HT increases during the sleeping stage.

These and other results raise many questions that are still unanswered. For example, does serotonin released from nerve cells initiate sleep by stimulating a specific receptor site, or does the mechanism that induces sleep depend on the serotonin concentration in the brain? What function does PS activity serve, and why do certain species not require it at all? Basic research could provide more data to determine if human insomnia can be treated by brain monoamine manipulation through drugs. An example would be a drug that would selectively increase cerebral serotonin, such as its precursor 5-HTP coupled to a peripheral dopa decarboxylase inhibitor that would permit penetration of the 5-HTP through the blood–brain barrier. These speculations will be resolved eventually. Even though some of the experimental evidence presented here is at times contradictory and incomplete, it still constitutes the most promising neurochemical correlate advanced thus far in this field.

3.8. BLOOD–BRAIN BARRIER

Systemic molecules traveling toward the brain encounter a series of barriers that can destroy, diminish, or inactivate them before they penetrate the brain substance. The term *blood–brain barrier* refers to the ability of cerebral vessels to selectively regulate the passage of molecules from the inner lining of the capillary walls to the most superficial layer of the brain parenchyma surrounding the tissue. It is also believed that this mechanism of selective permeability functions to allow discharge of particles from the cerebral tissue into the cerebral vessels for elimination. The process then is a two-way barrier system involving a certain exchange of fluids from the periphery to the ground substance and vice versa. Many substances such as electrolytes, colloids, and vital dyes do not penetrate the central nervous system from the bloodstream. Since the brain has an almost unsurpassed supply of arterial blood and consumes about 20 % of the oxygen in the whole body, a protective mechanism not only prevents the incorporation of large amounts of electrolytes and colloids but at the same time maintains the circulation through the brain of large amounts of arterial blood. The blood–brain barrier can be detected in the embryonic brain as well as in the brains of the lowest vertebrates. Even after death, at least for a few hours, it seems to survive after the protected brain itself ceases to function. The only way in which a general breakdown of the blood–brain barrier can be achieved *in vivo* is through drastic measures, and such a breakdown almost invariably leads to death.

Most of the data obtained on the blood–brain barrier up until the last decade were based on vital dye studies, to which fluorescent dye studies were added later. Although some knowledge was gained about the blood–brain barrier function using dyes, they did not necessarily reveal the normal physiological processes. For one thing, most of these dyes proved to be highly toxic to experimental animals. The molecules were of relatively large size and therefore gave no information on the exchange of ions through the barrier. The modern concept of the site of the blood–brain barrier places it in the inner membrane of the intracerebral capillaries, while the site of the blood–cerebrospinal fluid barrier is thought to be in the inner membrane of the capillaries of the choroid plexus and meninges. This brain barrier, therefore, would form a relationship between the capillary wall and the most superficial layer of the cerebral tissue surrounding the vessel. Thus, the anatomical basis of the blood–brain barrier should be sought in the relationship between the blood capillaries and the neurons.

There are two structures which appear to intervene between the capillaries and the neurons: (1) the ground substance, which is mucopolysaccharide in nature, and (2) the perivascular neuroglia. Morphological and ultrastructural studies of the central nervous system capillary regions indicate that the cells in the central nervous system are closely packed, leaving an area perhaps 150–200 Å wide where fluid material is found interlaced between cellular elements. This observation led Edström (1964) to suggest that such a narrow space would impede not only convection flows but also longitudinal diffusion between the cell surfaces. Transcellular transport modes may predominate in the central nervous system, and the perivascular end feet of the neuroglia could contain specific uptake or elimination mechanisms in the central nervous system capillary exchange. A great deal of information has appeared within recent years as to the blood–brain barrier mechanism for the monoamine precursors dopa and 5-hydroxytryptophan (5-HTP). The blood–brain barrier to these two precursors does not appear to be a physical barrier but rather a chemical one, dependent on the enzyme dopa decarboxylase (Fig. 30). Dopa decarboxylase is found within the capillaries along with other enzymes such as monoamine oxidase. The monoamines dopamine, noradrenaline, and serotonin do not appear to cross the blood–brain barrier in any significant amounts. The lipid solubility of substances crossing the blood–brain barrier may also have some bearing on the blood barrier mechanism. Noradrenaline, for example, being a highly polar substance, does not readily penetrate the brain barrier. If noradrenaline is given intraperitoneally, it exerts its effects mainly on peripheral receptors. In large doses, it may produce some excitement due to small amounts entering the brain. This would also seem to follow for adrenaline. It is well known that peripherally administered adrenaline produces mental and central neurophysiological effects as though acting

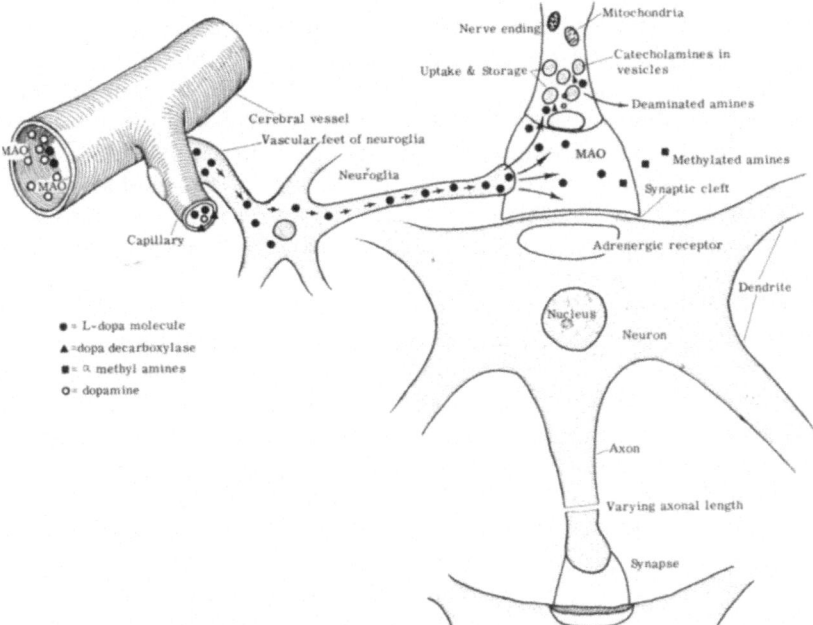

Fig. 30. Highly simplified representation of blood–brain barrier phenomenon after injection of L-dopa. Dopa molecules (●) that arrive in brain are quickly decarboxylated (▲) within the capillaries (endothelial cell layer) to dopamine (O); MAO then deaminates the remaining capillary dopamine. Some dopa molecules can cross the brain barrier if the enzyme dopa decarboxylase is peripherally saturated with dopa (chronically). Almost complete penetration of dopa from the capillary into brain tissue is achieved after peripheral dopa decarboxylase inhibition (MK-486, Ro 4-4602, etc.). Neuronal decarboxylation of L-dopa molecules occurs in the cytoplasm of the nerve ending (▲→O), resulting in dopamine. The latter molecules then penetrate storage vesicles, where hydroxylation produces noradrenaline. The homeostatic balance of the amines within the nerve cell is regulated by MAO, which deaminates dopamine.

directly on the central nervous system, although radioactive adrenaline infused into an animal's femoral vein is not found in the brain. This suggests that adrenaline does not penetrate the blood–brain barrier. The effects of adrenaline may be explained by its interaction with the hypothalamus or peripheral receptors. In the case of noradrenaline, lipid-soluble congeners of this amine penetrate the blood–brain barrier and cause stimulation of central adrenergic receptors.

When either of the precursors dopa or 5-HTP is injected into the living system, two things appear to happen: (1) biotransformation of dopa to dopamine, or 5-HTP to serotonin, and (2) oxidative deamination by monoamine oxidase at the peripheral level. This section will attempt to examine the process by which either dopa or 5-HTP is able to penetrate the brain barrier to enhance dopamine and serotonin levels in endogenous brain regions. Much of our present knowledge involving blood–brain barrier mechanisms for L-dopa and 5-HTP is the result of the development of the histochemical fluorescence technique in 1962 by Falck and his colleagues in Sweden. By use of this technique, dopamine-, noradrenaline-, and serotonin-containing terminals and cell bodies are visible as fluorescent fluorophors in brain tissue. The monoamine precursors dopa and 5-HTP are also visible after fluorescent processing, but since they are readily taken up in the free form in nerve tissue they are not normally confused with endogenous or bound monoamines in the brain.

The first evidence of a blood–brain barrier mechanism for dopa was presented by Bertler et al. (1963c, 1966). This was done by treating rats with a MAO inhibitor, nialamide, followed by a dose of L-dopa. The subsequent fluorescent picture showed a capillary network, resulting from the decarboxylation of dopa to dopamine at the peripheral level. Strongly fluorescent pericytes could be seen within the endothelial lining of brain capillaries. The newly transformed dopamine remained at the capillary level, not being able to penetrate the blood–brain barrier. Several years later, Constantinidis et al. (1968, 1969a, b) and de la Torre (1968b, 1970a, b) showed that gradual penetration of L-dopa into the brain tissue could be achieved by partially inhibiting peripheral dopa decarboxylase localized in the capillaries. It then became possible to study the relative penetration of L-dopa into selective brain regions as well as into the major organs (kidney, heart, spleen, lung, liver).

Subcortical capillary fluorescence following nialamide–L-dopa treatment was shown to be evenly distributed throughout all brain regions, although the capillary density was always stronger in the gray matter than in the white matter. If L-dopa was injected intraperitoneally without prior administration of a MAO inhibitor, a weak capillary fluorescence was seen in the anteroposterior length of the brain after about 30 min. This weak capillary fluorescence was due to rapid decarboxylation of the L-dopa into dopamine, which was subsequently deaminated by active MAO. However, if the enzyme MAO was inhibited prior to administration of L-dopa, the strong capillary fluorescence was visible for many hours after the death of the animal. If the dopa decarboxylase inhibitor [N-(DL-seryl)-N'-2,3,4-trihydroxybenzyl] hydrazine (Ro 4-4602) was injected after nialamide pretreatment but before L-dopa (Table 13), then penetration of dopa was possible due to the gradual inactivation of dopa decarboxylase at the peripheral level. The key to this study was use of very small doses of the Ro 4-4602 so that endogenous dopa decarboxylase activity would not be affected and only peripheral dopa decarboxylase would be inhibited by Ro 4-4602. It thus became possible to study by gradual increase of Ro 4-4602

Table 13.   Rat Brain Regions Showing Relative Dopa Penetration After Decarboxylase Inhibition

Brain regions[a]		Ro 4–4602 (mg/kg)[b]						
		0	2	4	6	8	10	50
Mammillary region	Pre and middle	x	x	x	x	x	0	0
	Lateral	x	x	x	x	x	x	0
Posterior hypothalamus		x	x	x	x	x	0	0
Ventromedial	Pars centralis	x	x	0	0	0	0	0
nucleus	Pars dorsalis	x	x	0	0	0	0	0
Dorsomedial	Pars dorsalis	x	x	x	x	x	0	0
nucleus	Pars ventralis	x	x	x	0	0	0	0
Arcuate		x	0	0	0	0	0	0
Periventricular	Inferior region (anterior)	x	x	x	0	0	0	0
hypothalamic	Inferior region (posterior)	x	x	0	0	0	0	0
nucleus	Superior region (anterior)	x	x	x	x	x	0	0
	Superior region (posterior)	x	x	x	x	x	0	0
Hypothalamohypo-	Lateral hypothalamic nucleus	x	x	x	x	x	0	0
physeal region	Recessus infundibularis	x	0	0	0	0	0	0
	Tuberoinfundibulum	0	0	0	0	0	0	0
	Anterior hypothalamic nucleus	x	x	x	x	x	x	0
Paraventricular	Pars parvocellularis	x	x	x	x	x	0	0
nucleus	Pars magnocellularis	x	x	x	x	x	x	x
Other hypothalamic	Supraoptic	x	x	x	x	x	x	x
nuclei	Suprachiasmatic	x	x	0	0	0	0	0
	Medial preoptic	x	x	x	x	x	x	0
	Lateral preoptic	x	x	x	x	x	x	0
Extrahypothalamic	Zona incerta	x	x	x	x	x	x	0
regions[c]	H$_1$ Forel's field	x	x	x	x	x	x	0
	H$_2$ Forel's field	x	x	x	x	x	0	0
	Neostriatum	x	x	x	x	x	0	0

[a] Hypothalamic regions (and neocortex) showing the relative decarboxylase activity for dopa penetration in the blood–brain barrier. Progressive doses of the dopa decarboxylase inhibitor Ro 4–4602 in Niamid (150 mg/kg) plus dopa (50 mg/kg) treated rats.

[b] Capillary fluorescence present (x) or absent (0) as compared to Niamid + dopa control (zero dose of Ro 4–4602). $N = 150$, $P < 0.001$.

[c] Included for comparison.

from 2 mg/kg to 50 mg/kg the penetration of L-dopa or 5-HTP into the brain tissue (Fig. 31). Curiously, it was found that L-dopa first began penetrating the brain parenchyma at various selective sites in the hypothalamus. These primary sites of penetration corresponded to the lowest capillary densities in the brain. The area least vascularized in the brain in man or rat is the ventromedial hypothalamic nucleus, and this was consequently the first site of entrance of L-dopa after inhibition of dopa decarboxylase by 4 mg/kg of Ro 4-4602. Subsequent penetration followed increased

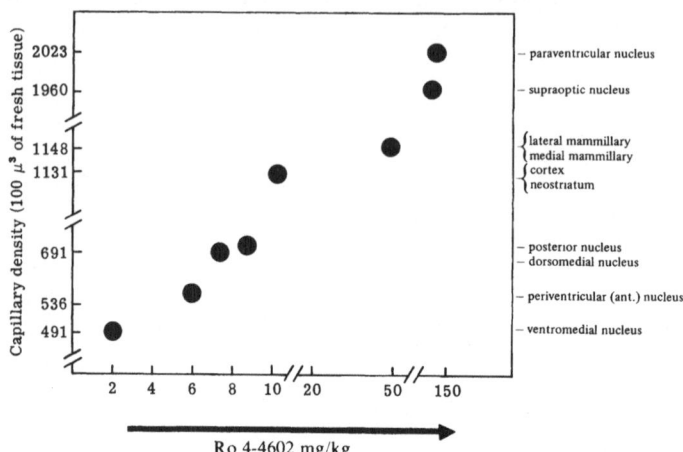

Fig. 31. The relationship between progressive brain peripheral decarboxylase inhibition by Ro 4-4602 and L-dopa penetration into various
cerebral structures. Cortex and neostriatum are included for comparison.
$N = 150$, $P < 0.001$. (After Craigie, 1924.)

capillary density, so that at 6 mg/kg of Ro 4-4602, L-dopa was seen to enter
the dorsomedial hypothalamic nucleus. Several general observations could
be made in regard to gradual or partial inhibition of the enzyme dopa
decarboxylase at the peripheral level. First, as already stated, dopa penetration in the brain tissue was dependent on the capillary density of that
of that region. Second, penetration seemed to favor the inferior half and the
lateral aspects of the hypothalamus, while areas of increased vascular
density such as found in the supraoptic nucleus and the paraventricular
nucleus of the hypothalamus were the last regions of dopa penetration.
After partial inhibition of dopa decarboxylase, not only did L-dopa penetrate into the regions already mentioned but there was also a decrease in the
intensity and numbers of fluorescent capillaries in such areas as the cingulate,
the entorhinal, and the amygdaloid cortex as well as in the thalamus,
midbrain, pons, medulla, and white matter. Complete penetration of dopa
was seen following 150 mg/kg of Ro 4-4602. At this dose, the brain sections
appeared as a mass of diffuse green fluorescence even though it was still
possible to discern some catecholamine-containing cell bodies and nerve
terminals.

Another aspect of nialamide plus L-dopa treatment was the physiological response of the animals to this combined treatment. Nialamide
injected at 150 mg/kg and L-dopa at 50 mg/kg caused a series of sympathetic
and parasympathetic reactions in the rat. These effects are summarized in
Table 14. The autonomic reactions could be decreased or completely

abolished by increasing dopa decarboxylase inhibition. Thus, after only 10 mg/kg of Ro 4-4602 in nialamide–dopa treated rats, some marked decreases in sympathetic and parasympathetic reactions were seen. After 50 mg/kg of Ro 4-4602, the animals' gross behavior appeared normal. Since no obvious physical symptoms were noted after nialamide alone or after dopa alone, the physical signs seen following nialamide–dopa treatment may have come about as a result of decarboxylation of dopa in the brain capillaries with massive accumulation of dopamine at that level. This suggests that the accumulated vascular dopamine exerts very strong pressor effects on the systemic circulation which may be proportional to the concentration of the monoamine in the cerebral vasculature. The vasopressor effects produced by dopamine may be counteracted by using a β-adrenergic blocker such as propranalol. This drug is also useful in patients with Parkinson's disease who are receiving massive doses of L-dopa. At therapeutic doses, propranalol appears to block the positive inotropic action of L-dopa. However, after 3 months of continuous L-dopa therapy there is apparent tolerance to the positive inotropic action. Intravenous administration of dopamine increases cardiac output, renal blood flow, and sodium excretion. The vascular effects of L-dopa administration will be discussed in Section 4.3.4.

Table 14. Sympathetic and Parasympathetic (Gross) Effects Seen in the Rat After Drug Treatment

Drugs	Sympathetic effects	Parasympathetic effects
Niamid + dopa[a]	Tachycardia, respiratory increase, pupillary dilation, piloerection, exophthalmos, Straub tail, hyperkinesia	Salivation, defecation, micturition, narrowing of palpebral fissure, occasional vomiting, stupor
Dopa[b]	None	None
Ro 4–4602 + dopa or Niamid + Ro 4–4602[c]	None	None
Niamid + dopa[d], 2–6 mg	Some progressive diminution of symptoms	Some progressive diminution of symptoms
When Ro 4–4602 follows Niamid at 10 mg	Tachycardia	Drowsiness
20 mg	None	None

[a] Central and autonomic effects observed after various treatments of dopa in MAO-inhibited rats.
[b] Effect of dopa alone.
[c] Combinations of Niamid, Ro 4–4602, and L-dopa.
[d] Reduction of effects correlates with increased dosage of Ro 4–4602 in Niamid + dopa rats (sequence: Niamid plus Ro 4–4602 at x mg/kg + dopa). Doses in the others were standard: dopa, 50 mg/kg; Niamid, 150 mg/kg.

Another method by which dopa penetration across the blood–brain barrier may be facilitated is by the use of a transport carrier, *viz.*, dimethylsulfoxide (DMSO). When L-dopa is mixed with DMSO in 1:10 vol and intraperitoneally injected, partial penetration of the amine is evident in the nerve tissue $\frac{1}{2}$ hr afterward. When the same experiment is repeated using 5-HTP, no penetration into the nerve tissue is seen. This penetration of L-dopa through the blood–brain barrier when in solution with DMSO might be interpreted as evidence that DMSO acts as weak decarboxylase inhibitor. However, this idea must be discarded in view of the following evidence: Since 5-HTP and L-dopa are both decarboxylated by the same enzyme, dopa decarboxylase, it is logical that either amino acid dissolved in DMSO should penetrate into the brain parenchyma, but 5-HTP does not. Also, if DMSO is given separately from L-dopa 1 hr before, no diffusion or penetration of L-dopa is seen in the brain parenchyma and the capillary picture resembles that of animals treated with nialamide plus L-dopa. Therefore, DMSO does not appear to weaken or inhibit the activity of dopa decarboxylase. Another possibility might be a chemical reaction between DMSO and L-dopa that gives a dopa analogue capable of partially crossing the blood–brain barrier. However, the following observation appear to contradict this idea:

(1) After L-dopa penetrates the brain parenchyma, there is an increase in the diffuse neostriatal fluorescence. This is seen after DMSO plus L-dopa injection. The reason for the increase in fluorescence in this region is probably the large amount of dopa decarboxylase concentrated in this area, which takes up the free inbound dopa and converts it into dopamine.

(2) A visible increase in the tuberoinfundibular fluorescence develops after L-dopa administration. This increase is also seen after DMSO plus L-dopa treatment. This brain region is, of course, outside the blood–brain barrier, and any free dopa in the area tends to potentiate the primary normal fluorescence. Furthermore, dopa analogues do not normally fluoresce in tissue when this technique is used.

(3) DMSO itself is a relatively intert chemical compound, so it is not likely that it would react with another substance. The most cogent explanation for the penetration of L-dopa mixed with DMSO is that DMSO is able to transport the amine across the lipid phase of the blood–brain barrier to some extent. This carrier activity is probably rapid, since any reasonable amount of L-dopa in the brain capillaries is quickly converted by dopa decarboxylase into dopamine, and this substance, of course, does not cross the blood–brain barrier.

What is not yet known is by what specific mechanism DMSO is able to transport the dopa molecules across the cerebral vasculature into the brain parenchyma. It has been shown, however, that DMSO is also able to aid the penetration of substances such as pemoline hydroxide through this brain barrier membrane, while other substances such as sugar, mixed with dimethylsulfoxide in the same manner, are not able to penetrate into the brain tissue (Brink and Stein, 1967). Certain brain lesions such as those produced by electrolytic or mechanical means sometimes may damage the blood–brain barrier permeability to various substances injected afterward. Johansson *et al.* (1970) have reported that acute hypertension produced by intravenous injections of metaraminol bitartrate (a close analogue of noradrenaline) can facilitate the penetration of Evans blue tracer into the brain tissue of cats. Extravasation of the Evans blue was evident about 10 min after a sudden increase in systolic pressure to 90 mm Hg induced by the metaraminol. Microscopic examination of the brain tissue revealed a breakdown of the blood–brain barrier to the tracer after the hypertensive insult. Björklund *et al.* (1969), using the histochemical fluorescence technique, found that following electrolytic lesions in the brain parenchyma of the rat, the intracerebral capillaries within a short distance from the lesion did not become fluorescent upon the systemic administration of L-dopa. However, 3–4 weeks after such a lesion, the new capillaries formed at the periphery of the lesions were able to trap L-dopa as before. It is possible then that after traumatic head injury, the blood–brain barrier may become more permeable to substances which do not normally penetrate into the brain parenchyma, possibly causing various neurological deficits such as seen in brain edema, convulsions, electroencephalographic changes, paralysis, or death.

Another recent finding along this line is worth mentioning. We had always wondered, for example, why it was in the early 1960s that Barbeau, Ehringer and Hornykiewicz, and several others who had tried acute administration of L-dopa in humans could only report relative and transient success with parkinsonian patients, even when pathological studies indicated a deficiency of dopamine in nigroneostriatal structures. It has been shown (de la Torre and Mullan, 1971, 1972; de la Torre and Boggan 1971) that acutely administered L-dopa in animals simply does not penetrate the blood–brain barrier in any significant amount and remains within the capillaries in the endothelial layer, where peripheral dopa decarboxylase enzyme quickly converts it to dopamine. Thus, the newly formed dopamine remains within the capillaries, unable to penetrate into the brain tissue due to its inability to cross the barrier. We believe that the failure in the early 1960s to ameliorate the symptoms of parkinsonian patients who received L-dopa orally and intravenously, never for more than 2 weeks at a time, was due to the fact that only very small amounts of the L-dopa entered the endogenous brain tissue, thus failing to increase central dopamine levels.

In order to substantiate this statement, a review of previous work is necessary. It is often quoted in the literature that L-dopa is able to penetrate into the brain tissue without difficulty after its administration. This was, and still is, accepted by many people in the field. In support of this contention, they cite various reports (Carlsson and Hillarp, 1962; Gey and Pletscher, 1964; Burack and Draskoczy, 1964; Persson and Waldeck, 1968) concerning the transformation of administered L-dopa to dopamine in various organs, including the brain. Although these papers (with the exception of that of Persson and Waldeck, 1968) do not imply any facility of the L-dopa for ready penetration into the brain tissue after its administration, nevertheless other investigators have concluded from the premise that L-dopa, whether or not it is labeled, can be decarboxylated to dopamine in the brain and that a rise in dopamine levels does take place in endogenous brain tissue. Two more findings also seem to support easy barrier penetration by L-dopa: successful L-dopa therapy in patients with Parkinson's disease, and psychophysiological manifestations (*viz.,* sham rage, excitement) generally observed in animals after heavy acute doses of L-dopa. The latter is attributed to the release of noradrenaline from adrenergic neurons in the brain induced by the newly synthesized dopamine and noradrenaline from the L-dopa.

We suggest that all these conclusions are erroneous. It is true that after the administration of L-dopa there is conversion to dopamine in the extracerebral organs and that the dopamine then proceeds through its hydroxylation process and is in turn transformed into noradrenaline. In the brain, however, this situation is markedly different. As we have discussed in the preceding pages, histochemical fluorescence shows that acutely administered L-dopa enters the endothelial lining of brain capillaries, where it is rapidly decarboxylated to dopamine. The exact site of this decarboxylation within the endothelial layer is as yet unknown but may be presumed to be within the endothelial cells. The evidence for this is that in ultrastructural studies of normal capillaries the cytoplasm of these endothelial cells usually protrudes into the capillary lumen. This phenomenon would explain the bulging or "balloon-like" protrusion of the green fluorescence extending from the endothelial cell layer of the brain capillaries into the lumen (Fig. 32). However, one thing appears certain: if L-dopa were able to penetrate into the brain tissue from the capillaries, then some amount of diffuse fluorescence around these capillaries should be visualized, as, for instance, is the case after minimal peripheral dopa decarboxylase blocking by an inhibitor of this enzyme in the hypothalamus. In the hypothalamus, after 2 mg/kg of Ro 4-4602, a slight green fluorescent diffusion is seen around some of the capillaries surrounding the ventromedial hypothalamic nucleus, while the brain tissue capillaries surrounding the dorsomedial nucleus, a few millimeters laterally, remain dark, indicating no penetration of the amino acid. Consequently, it is obvious that measurement of dopamine levels after brain homogenization of animals that have been treated with L-dopa results in the peripheral (vascular brain tissue) increase in dopamine

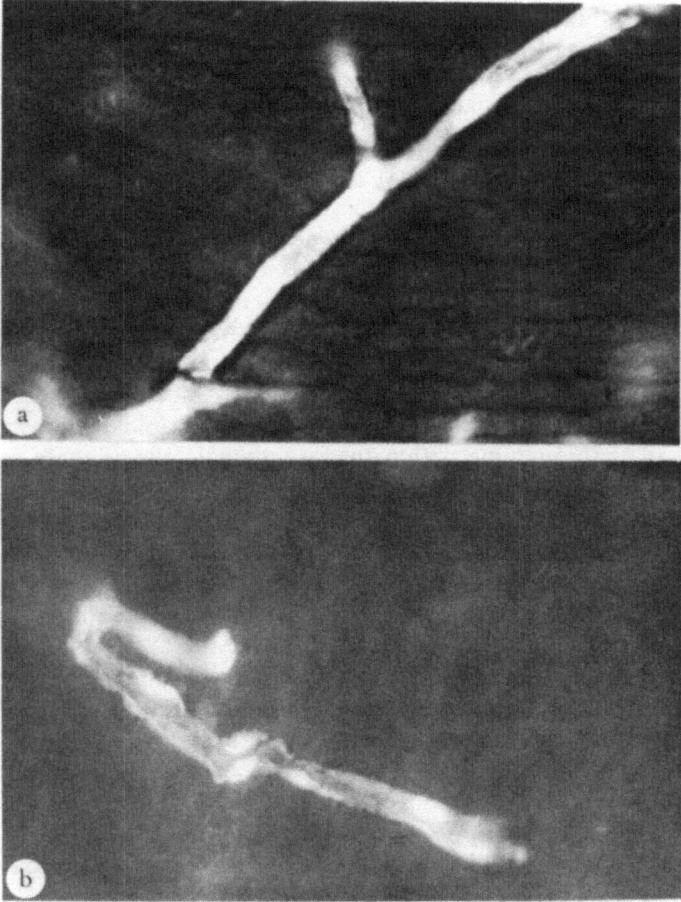

Fig. 32. Brain capillary lumen and endothelial cell fluorescence after L-dopa (50 mg/kg) in rat (a). Following cerebral perfusion with saline, fluorescence in lumen disappears leaving only endothelial cell layer fluorescence (balloon-like swelling on capillary wall) (b). Surrounding neuronal tissue in both animals receiving L-dopa remains dark. See explanation in text. Anterior thalamic regions. 190×. (Reduced for reproduction 20%.)

being measured as part of the endogenous (neuronal brain tissue) dopamine concentration. Thus, the separation procedure used in the biochemical assays can extract the trapped dopamine within the endothelial cell linings, and the total measured concentration of brain dopamine is thereby seen as being increased (Fig. 33).

Furthermore, in biochemical assays there is always a slight but consisten increase in the brain noradrenaline levels. But if the trapped dopamine in the endothelial lining of the capillaries is not being hydroxylated to noradrenaline and the L-dopa is not penetrating into the brain tissue, how then to account for this slight increase in brain noradrenaline levels? This, we believe, may be due to circulating noradrenaline which is hydroxylated in extracerebral organs (or in blood serum; see Weinshilboum and Axelrod, 1971) after about 1 hr following the L-dopa administration and is in the capillary lumen at the time the brain is homogenized and the amines are extracted. This would explain Persson and Waldeck's (1968) report that after labeled L-dopa is injected into mice, an increase in labeled noradrenaline occurs during the first 4 hr but not at the expense of an equimolar amount of labeled dopamine in the brain. It is reasonable to assume that as hydroxylation of labeled dopamine occurs extracerebrally, the amount of circulating labeled noradrenaline in the brain capillary lumen is independent of the dopamine within the endothelial cells of these capillaries formed after the decarboxylation of injected L-dopa-H^3. The rates of conversion of administered labeled dopa to dopamine and noradrenaline also appear to be different (Burack and Draskoczy, 1964) in extracerebral organs such as the heart and the spleen (there is considerably more labeled dopamine in the spleen than in the heart 2 hr after the administration of L-dopa-H^3). Moreover, it would seem that if the tagged L-dopa were to pass through the blood–brain barrier to increase dopamine levels endogenously, then such increased stored dopamine would remain viable within its storage granule, protected from enzymatic breakdown. There would also be increased radioactive noradrenaline within these granules which would remain safely protected from enzymatic degradation until its release. However, this does not appear to be what happens, since the radioactivity of L-dopa, dopamine, and noradrenaline drops to zero (Gey and Pletscher, 1964) about 9 hr after its maximum activity at 20 min. This indicates that the labeled dopamine within the brain capillaries is deaminated by active monoamine oxidase, probably located in the mitochondrial fraction within the capillary endothelium, and that the circulatory labeled noradrenaline is also degraded by the COMT in red blood cells (Cohn et al., 1971) or by monoamine oxidase in extracerebral organs. We offer the following additional evidence: Intracardiac left ventricular perfusion of the brain (de la Torre and Boggan, 1971) shows that 30 min following intraperitoneal injections of L-dopa in rats, 50 mg/kg, there is a 42 % decrease in the levels of brain dopamine when measured by biochemical assays as compared to sham-perfused animals (see p. 97). Moreover, peak brain dopamine values are observed at about 30 min following L-dopa injection as measured by biochemical assays. After 30 min, the cerebral dopamine values decrease rapidly. This indicates to us that by perfusing the cerebral vasculature with saline after L-dopa administra-

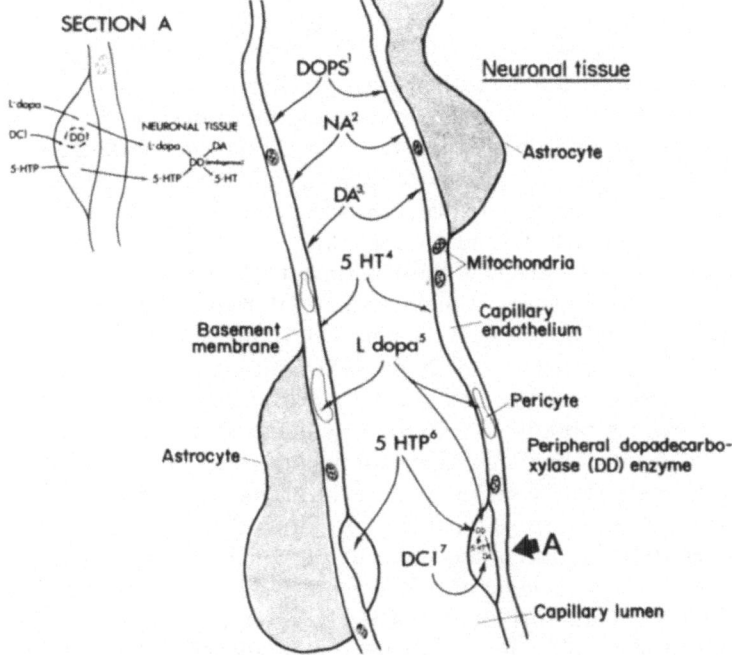

Fig. 33. Longitudinal section of a brain capillary showing the fate of monoamines, their precursors, and peripheral decarboxylase inhibitors after acute
administration in rats. Evidence suggests that a comparable mechanism
exists for higher mammalian brains, including that of man. Compounds
such as DOPS, NA, DA, and 5-HT do not enter the endothelial cell layer
of the capillaries but merely circulate within the lumen. Monoamine precursors (L-dopa, 5-HTP), DCI such as mentioned in the text, and the monoamine oxidase inhibitor nialamide can apparently enter the endothelial
cell layer. L-Dopa and 5-HTP are decarboxylated within this layer by DD.
The newly transformed DA and 5-HT appear to accumulate within the
pericytes and endothelial cells (section A) and remain there until destruction
by monoamine oxidase, possibly located in the mitochondria. If, however,
DCI is given prior to L-dopa or 5-HTP, then DD is blocked and the precursors are able to pass through the basement membrane, where they are
taken up by astrocytes and conveyed to neuronal tissue, where endogenous
DD converts them to DA and 5-HT (section A). The hypothetical model
proposed would explain why acute doses of amine precursors do not drama-
tically alter the symptoms of Parkinson's disease and why brain vascular
perfusion is able to decrease total brain DA content after L-dopa as well as
other phenomenological aspects of blood–brain barrier mechanisms for these
substances (see text). Magnification of model (from electron microscopic
photographs) $42,800\times$. (Reduced for reproduction 50%.) Key: (1) DOPS,
dihydroxyphenylserine; (2) NA, noradrenaline; (3) DA, 3,4-dihydroxy-
phenylethylamine; (4) 5-HT, serotonin; (5) L-dopa, 3,4-dihydroxyphenyl-
alanine; (6) 5-HTP, 5-hydroxytryptophan; (7) DCI, peripheral dopa decarboxylase inhibitor; DD, peripheral and endogenous dopa decarboxylase
enzyme.

tion, one may markedly alter the total brain dopamine concentrations expected to increase. We believe this may occur because the L-dopa is rapidly decarboxylated in extracerebral organs, producing circulatory dopamine which is then washed off from the lumen of brain capillaries. Histochemical fluorescence data confirm disappearance of the green fluorescence in the *lumen* after perfusion. We cannot decrease the total dopamine concentration in the brain to normal nontreated values because the saline perfuses only the inner or luminal surface and cannot extract the trapped dopamine within the endothelial capillary cells. If the experiment is repeated and the animal is perfused after 15 min, the reduction in total brain dopamine levels as compared to the sham-perfused animals is only 14%. After 1 hr, the difference between the perfused and sham-perfused animals is only about 4%. It seems obvious, then, that the circulatory dopamine formed from extracerebral organs is rapidly destroyed; it is detected in normal concentrations several hours following L-dopa administration.

This brings us to the second part of the argument concerning the relief of symptoms in Parkinson's disease after L-dopa therapy. Clinical trials by Duvoisin *et al.* (1969) and others indicate that amelioration of symptoms in Parkinson's disease is evident only after many weeks of therapy and in doses averaging 6 g per day per patient. Clearly, then, acute and low doses of L-dopa alone are useless for treatment of this disorder. How the L-dopa finally leaks through the blood–brain barrier after saturating the system for many weeks in parkinsonian patients cannot be explained at the present. Perhaps peripheral dopa decarboxylase is somehow inactivated by the massive administration of L-dopa, which then permits the amino acid to penetrate into the brain tissue. Perhaps it is a failure of the blood–brain barrier mechanism resulting from the increased accumulation of L-dopa in brain capillaries. This might also explain some of the undesirable side effects in these patients seen after long-term L-dopa administration. If a peripheral dopa decarboxylase inhibitor is used prior to the L-dopa administration, the amelioration of the parkinsonian symptoms is seen within a few days instead of weeks (Tissot *et al.*, 1969; Siegfried *et al.*, 1969). In addition, the side effects are greatly diminished or nonexistent. More important, the effects of such a combined drug treatment are evident after only 700–900 mg of L-dopa.

We will not dwell at length on the third argument, which concerns the psychophysiological manifestations of L-dopa administration in animals. These effects are probably not central but peripheral (Reis *et al.*, 1970; Randrup and Munkvad, 1969). We can cite as an example the behavioral alterations produced by adrenaline in both animals and man (Weil-Malherbe *et al.*, 1959; Marañon, 1924), which indicate that although adrenaline does not penetrate the blood–brain barrier, it may nevertheless exert some effects on peripheral receptors which result in the various mood and

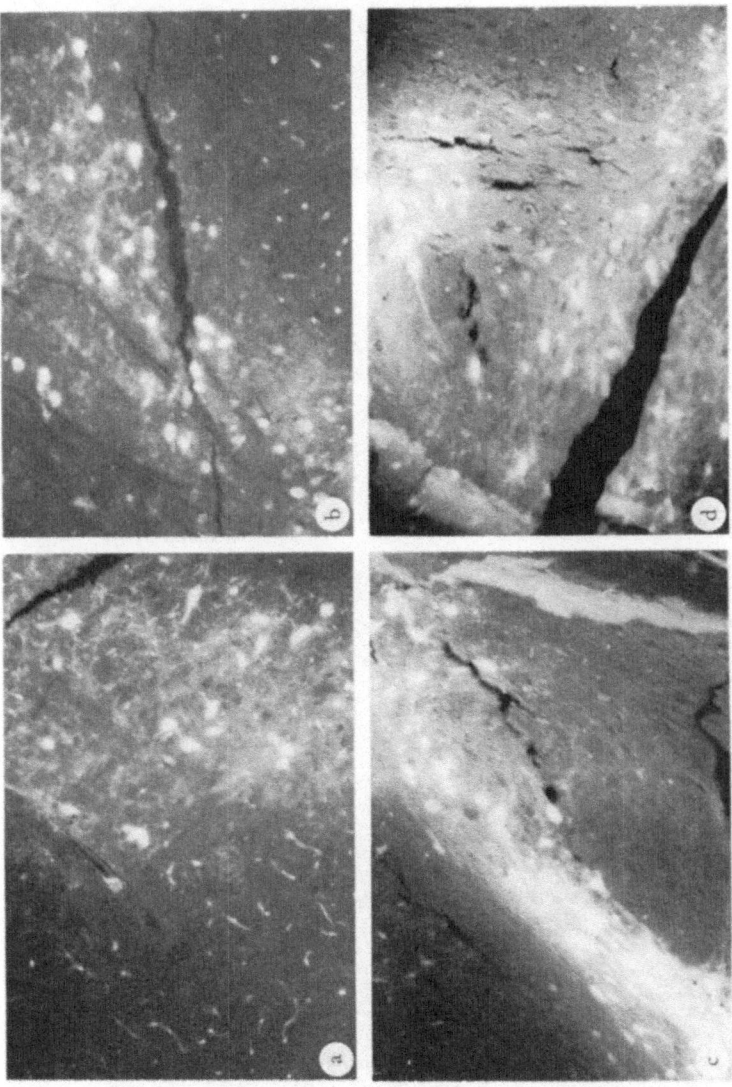

Fig. 34. Progressive penetration of L-dopa through the blood–brain barrier by prior treatment with MK-486. Rat substantia nigra, pars compacta. 90 ×. L-Dopa (50 mg/kg). (a) L-Dopa only; nigral neurons surrounded by strong capillary fluorescence. (b) MK-486 (12 mg/kg) plus L-dopa: some decrease in capillary fluorescence. (c) MK-486 (16 mg/kg) plus L-dopa: increase of nigral neuron fluorescence intensity. A few capillaries are still seen (top left). (d) MK-486 (24 mg/kg) plus L-dopa: diffuse fluorescence and increase in intensity of nigral neurons. No capillaries. (Reproduced for reproduction 30%.)

102 Chapter 3

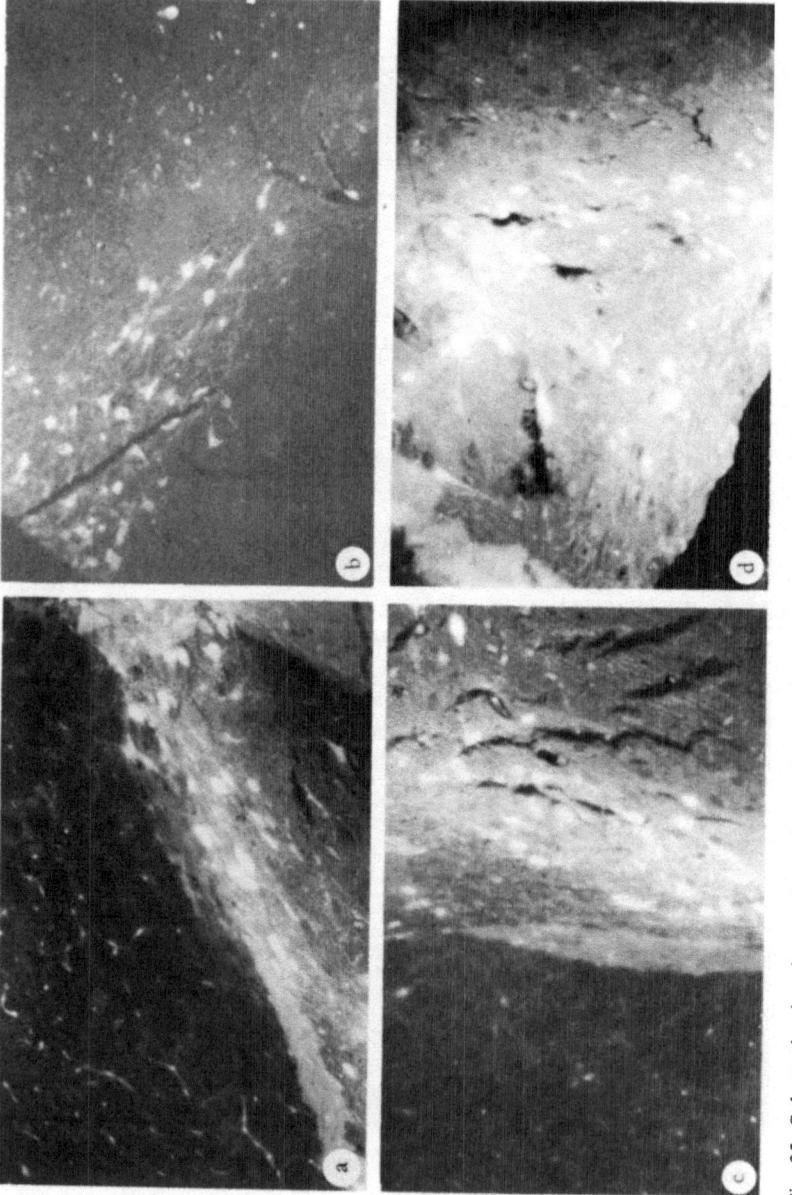

Fig. 35. Substantia nigra in rat after L-dopa (50 mg/kg). Same as Fig. 34 but with Ro 4-4602. (a) L-Dopa only. (b) Ro 4-4602 (2 mg/kg) plus L-dopa: decrease in capillary fluorescence. (c) Ro 4-4602 (4 mg/kg) plus L-dopa: increase in nigral neuron fluorescence. Faint capillaries present (left). (d) Ro 4-4602 (10 mg/kg) plus L-dopa: total disappearance of capillaries and strong neuronal and parenchymal fluorescence.

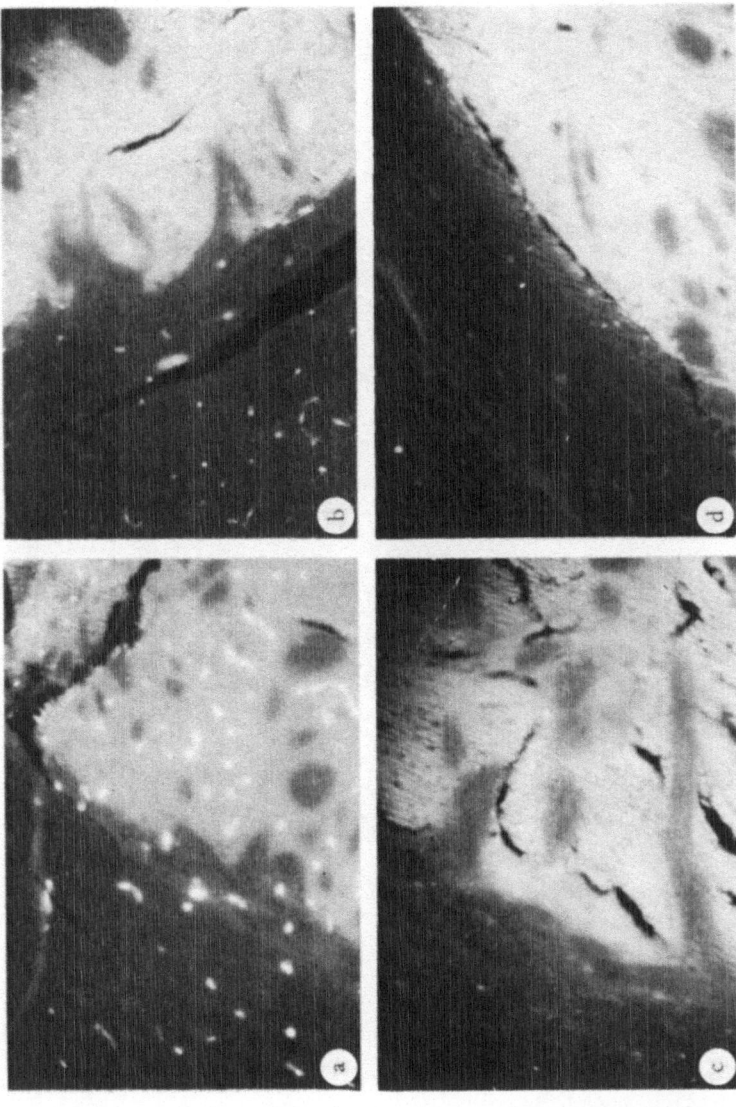

Fig. 36. Progressive penetration of L-dopa through the blood–brain barrier by prior treatment with MK-486. Rat caudate nucleus. L-Dopa (50 mg/kg). 90 ×. (a) L-Dopa only. Strong capillary fluorescence in caudate and radiation of corpus callosum (left). (b) MK-486 (8 mg/kg) plus L-dopa: increase in intensity of caudate fluorescence, some capillaries still visible. (c) MK-486 (10 mg/kg) plus L-dopa: very few, faint capillaries. (d) MK-486 (12 mg/kg) plus L-dopa: total disappearance of capillary fluorescence. Strong, diffuse caudate fluorescence.

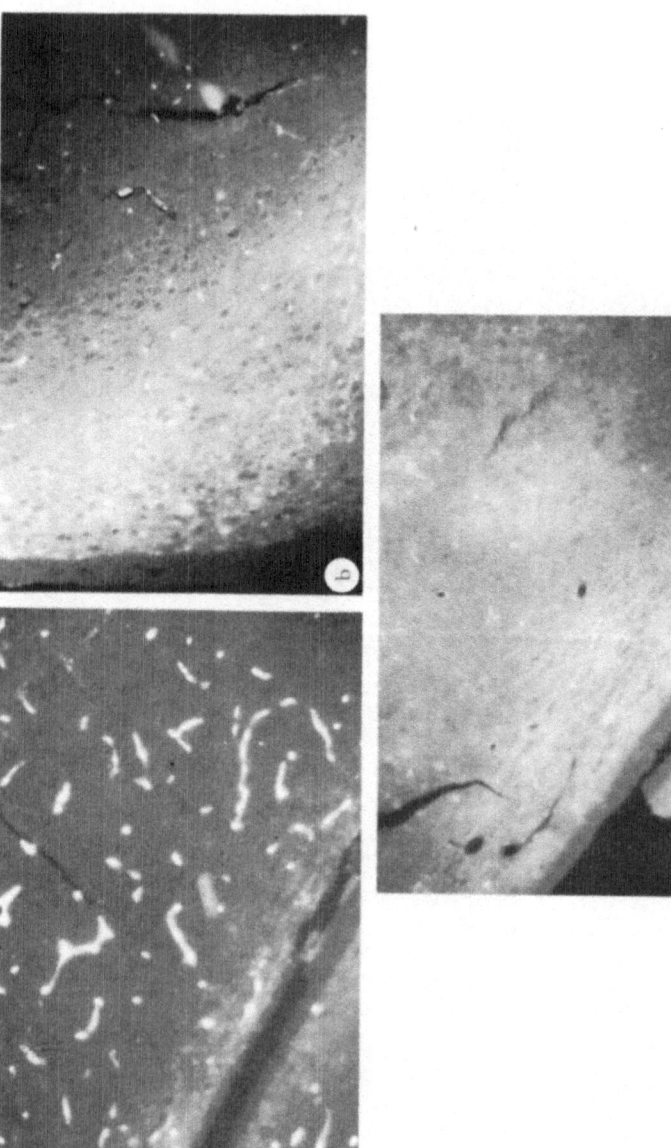

Fig. 37. Progressive penetration of L-dopa with a peripheral decarboxylase inhibitor (MK-486) through the blood–brain barrier at the level of the periventricular hypothalamic region. Rat third ventricle. (a) L-Dopa (50 mg/kg): strong capillary fluorescence, neuronal tissue dark, no penetration. 90 ×. (b) MK-486 (6 mg/kg) plus L-dopa (50 mg/kg): partial penetration extending from ventromedial hypothalamic nucleus (bottom of photograph) to ependymal wall (left) of third ventricle. Note that some capillaries are still seen near dorsomedial nucleus (right). 90 ×. (c) MK-486 (12 mg/kg) plus L-dopa (50 mg/kg): almost total penetration of L-dopa is now marked by lack of capillary fluorescence, which is replaced by intense, diffuse fluorescence due to free L-dopa. 90 ×. (Parts a–c reduced for reproduction 30%.)

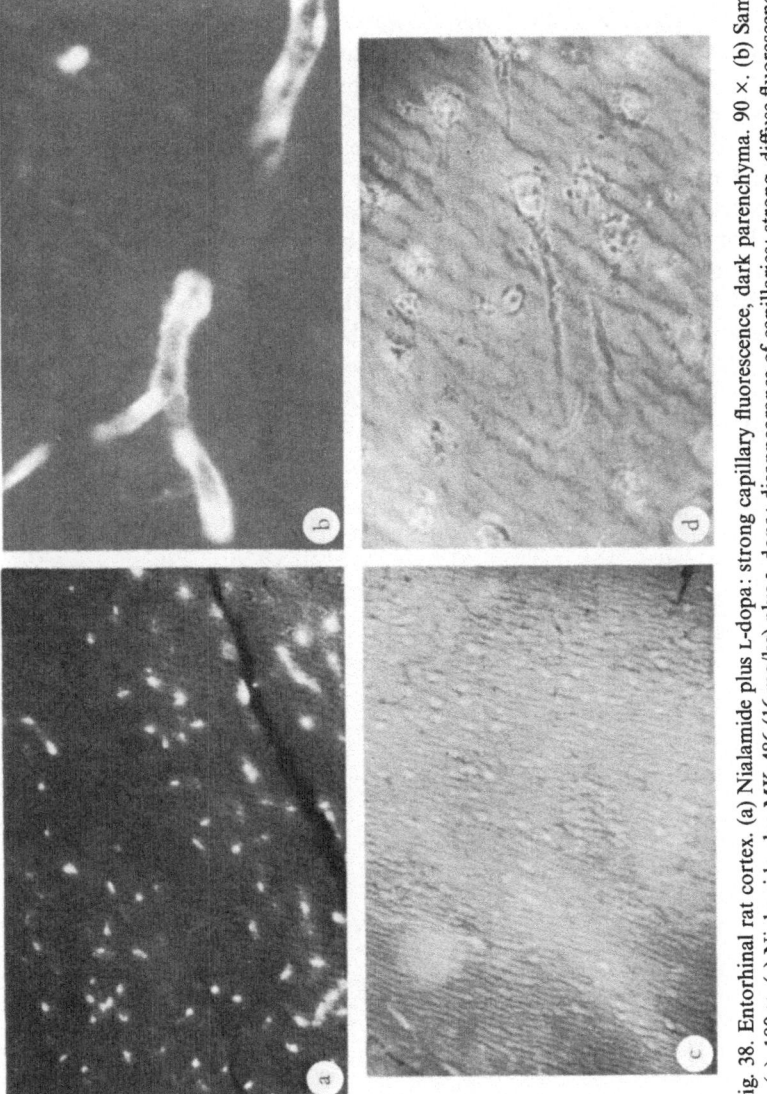

Fig. 38. Entorhinal rat cortex. (a) Nialamide plus L-dopa: strong capillary fluorescence, dark parenchyma. 90 ×. (b) Same as (a). 180 ×. (c) Nialamide plus MK-486 (16 mg/kg) plus L-dopa: disappearance of capillaries; strong, diffuse fluorescence due to free L-dopa in neuronal tissue. 90 ×. (d) Same as (c). 180 ×. Cortical neurons (nonmonoaminergic) are visible due to intense neuronal tissue illumination. 180 ×. (Parts a–d reduced for reproduction 25%.)

behavior changes observed. Besides this, our experiments show that no gross behavioral changes are observed in rats pretreated with a peripheral decarboxylase inhibitor and L-dopa (de la Torre, 1971).

One way of avoiding the barrier mechanism for the monoamines is through intraventricular injections. However, the deposition of monoamines can only be studied in limited topographical areas surrounding the ventricles. Nevertheless, the discovery and subsequent studies of the enzymic brain barrier mechanisms for L-dopa and 5-HTP (Figs. 34–38) have indicated for the first time a method of approach toward understanding this complex barrier system. Knowledge of the blood–brain barrier to monoamine precursors has opened a field of research of unprecedented possibilities. For example, considering the possible neurotransmitter role of the monoamines, if we can manipulate their concentrations we can then approach by neuropharmacological means the study of the very substrate behavior and thought. This is predicted to become one of the most exciting fields in the study of brain research. Present knowledge is already being applied in limited experimental human trials, which will be discussed in Section 4.3.

Neuropsychopathology

4.1. GENETIC FACTORS

4.1.1. Phenylketonuria (Phenylpyruvic Oligophrenia)

Phenylketonuria (PKU) is an inborn error in the metabolism of phenylalanine, which leads to idiocy in early childhood. About 25% of all PKU patients suffer from epileptic seizures and minor neurological symptoms such as tremors, ataxia, and hypertonicity of muscles. The basic biochemical defect is lack of the hydroxylating enzyme responsible for the conversion of phenylalanine to tyrosine in the liver. As a result, a high plasma concentration of phenylalanine occurs along with several abnormal metabolites; these are excreted by PKU patients as phenylpyruvic acid, a deaminated product of excess phenylalanine in plasma. A correlation between PKU and abnormal indole metabolism is also evident. This is characterized by a decrease in 5-hydroxyindoleacetic acid urinary levels and by an increased urinary excretion of indole-3-acetic acid. Hsia *et al.* (1964) have suggested three possibilities to explain PKU: (1) excessive phenylalanine might inhibit the hydroxylation of tryptophan metabolism in brain or liver, thus reducing available 5-HTP; (2) this excess might also inhibit the active transport of tryptophan or 5-HTP across the blood–brain barrier or brain cell membrane; and (3) excess phenylalanine and its metabolites might inhibit the decarboxylation of 5-HTP.

Studies show, however, that the excessive levels of phenylalanine and tyrosine in blood have no effect on the 5-HTP decarboxylase or monoamine oxidase in the brain. In addition, the increased tyrosine in PKU patients does not appear to interfere with tryptophan. These findings have led Hsia *et al.* to conclude that diminished serotonin in experimental PKU results from interference in the active transport of either tryptophan or 5-HTP across the blood–brain barrier. Woolley and Barthwal (1964) also found that mice in which PKU was experimentally induced by phenylalanine plus tyrosine administration had subnormal "maze learning" ability. In mice, this mental inadequacy could be prevented by continuous administration of serotonin analogues from birth to maturity.

Some of the more common symptoms that have been described for PKU are (1) abnormal movements of the fingers resembling athetosis, (2) ash blonde hair and eyebrows, (3) flexion of the arms and pointed head. As a result of numerous studies it has been shown that PKU is a hereditary disease provoked by a recessive autosomal gene. The psychopathology associated with PKU may be related to toxic concentrations of phenylpyruvic acid reaching the brain. Berger (1962) has postulated that the cerebral pathology seen in PKU is the result of complex metabolic disturbances provoked by the accumulation of phenylalanine unable to proceed to tyrosine. This excess phenylalanine may disturb the cellular chemistry through mechanisms regulating inhibition and adaptation. The end result would be slowing down of myelinization to produce a dysfunction in the cells of the extrapyramidal nuclei and of the cerebral cortex, resulting in extrapyramidal diseases and mental retardation. The pigmentation symptom may be due to inhibition of the enzyme tyrosinase caused by the increased levels of phenylanine and by the lack of tyrosine, which in turn lead to a deficiency in melanin production. The high levels of phenylalanine and its keto derivatives in the living system could affect the metabolism of tryptophan. In effect, there is a decrease in blood serotonin and in its urinary metabolite 5-hydroxyindoleacetic acid concomitant with an increase in indole derivatives. Either the decrease in serotonin or the presence of indole compounds could very well affect the cerebral function of the patient with PKU, resulting in the symptoms mentioned. Our understanding of this metabolic disturbance, that is, the accumulation of phenylalanine and its derivatives, has made possible effective treatment of these patients, consisting of a diet rigidly regulated to exclude any amounts of phenylalanine. The diagnostic screening for PKU of all newborn children can effectively eliminate this disease. Knox (1960) reports that in 43 PKU patients treated with a diet deficient in phenylalanine before the age of 6 months there did not appear to occur any brain pathology or any progression of the symptoms related to PKU.

4.1.2. Hepatolenticular Degeneration (Wilson's Disease)

Hepatolenticular degeneration (Wilson's disease) is characterized in late childhood or early adolescence by progressive damage to the nervous system, resulting in such symptoms as dystonia, tremor, rigidity, mental retardation, and behavioral disorders leading to psychotic states. Abnormalities in catecholamine and indole metabolism have been reported by Sourkes *et al.* (1963), who found increased urinary levels of dopamine and adrenaline in subjects with this condition. Sunderman (1963) also observed increased urinary excretion of 5-hydroxyindoleacetic acid, tryptamine, and other indole metabolites. Wilson's disease is also characterized by extrapyramidal disorders and by the presence of Kayser–Fleischer pericorneal ring (due to a deposition of copper granules in Descemet's membrane) as well as by mental retardation. Deposits of copper are found throughout the nervous system, but lesions are predominant in the putamen and caudate nucleus. These basal ganglia structures have a spongy aspect (bilaterally), and microscopically there is some loss of nerve cells and myelinated fibers from both nuclei. Multiplication of astrocytes and nuclear enlargement are seen, particularly in the putamen. (These astrocytes are called *Alzheimer type II cells.*) To the best of our knowledge, L-dopa treatment has never been tried at the onset of the disorder.

4.1.3. Hartnup Disease

The clinical manifestations of Hartnup disease are regular symptoms including cerebellar ataxia, pellagra-like rashes, and psychotic states characterized by behavioral disorder. According to Milne *et al.* (1960), the condition is most likely to appear in childhood and unlike similar diseases seems to improve with increasing age. The most constant biochemical finding is a generalized amino aciduria with metabolic blocking of the tryptophan metabolite kynurenin. In addition, there is an increase in the rate of urinary excretion of indole-3-acetic acid, indican, and indole-3-acetoglutamine, indicating a deficiency of tryptophan pyrrolase. PKU, Wilson's disease, and Hartnup disease are all probably inherited through recessive genes. In PKU, transmission occurs as the result of a single autosomal recessive gene carried by the parents, which explains the absence of clinical symptoms in the parents studied. As much as 200 mg of indoleacetic acid can be excreted in the urine of patients with Hartnup disease. These and other findings suggest that a defect in the metabolism of tryptophan is present. Since kynurenin is excreted in subnormal amounts and because the symptoms resemble pellagra, some deficiency in the pathway from tryptophan to nicotinic acid is probable; however, dietary administration of nicotinic acid does not alter the course of the disease. Nicotinamide therapy causes only a slight improvement of the clinical picture but does not even appear to correct the amino acid transport defect. It should be noted that urinary excretion of other amino acids such as pyrroline, serine, asparagine, and glutamine is increased. One problem in studying Hartnup disease is that many of its features are still unknown, since only a few cases have been described so far.

4.1.4. Mongolism (Down's Syndrome)

Evidence for the participation of monoamines and their precursors in other genetic diseases affecting the nervous system is still lacking. Tissot *et al.* (1966) studied the urine of 18 patients with mongolism (Down's syndrome). They found a decrease in the urinary excretion of 5-hydroxy-indoleacetic acid, especially in adults. A reduction in urinary excretion of xanthurenic acid, a direct metabolite of tryptophan, was the most consistent finding.

A reversal in the hypotonia of infants with Down's syndrome has been reported by Bazelon *et al.* (1967). This group found that, due to unknown causes, a depression of whole blood serotonin occurred in Down's syndrome of type trisomy-21. This abnormality was also detected in neonatal mongoloids. The reversal of the hypotonia was accomplished by giving 14 infant mongoloids between 0.15 and 0.5 mg/kg of 5-HTP by mouth for 1–7 weeks. Muscle tone was seen to improve on the second day of treatment. The 5-hydroxyindoles in whole blood increased gradually after 5-HTP treatment from 90 mg/100 ml to 130 mg/100 ml (normal range 120–230 mg/100 ml). It is possible that whole blood depression of 5-HT could reflect defective indoleamine metabolism in the central nervous system, particularly since serotonin acts to stimulate all smooth muscles. It will be of interest to gain some additional supportive evidence on the value of 5-HTP therapy in this disorder.

4.1.5. Familial Dysautonomia (Riley–Day Syndrome)

Familial dysautonomia (Riley–Day syndrome) is a rare congenital condition with special disturbances of the nervous system characterized by emotional instability, lack of sensitivity to pain, deformation of the spinal cord (scoliosis), hypertension, motor incoordination, and profuse sweating. The syndrome occurs primarily in children of European Jewish ancestry, but it may also be seen in adults.

Gitlow *et al.* (1970) measured the urinary excretion products in 82 dysautonomic patients and found marked elevations of homovanillic acid (HVA) as well as a reduction in vanillylmandelic acid and 3-methoxy-4-hydroxyphenylethyleneglycol, all of which are catecholamine metabolites. It was the conclusion of these workers that dysautonomia may involve a defect in the synthesis of noradrenaline. The evidence cited for this is the increase in the dopamine metabolite HVA excreted by these patients. This is an interesting hypothesis in view of the motor disturbances evident in these subjects. If noradrenaline synthesis is impaired in the central nervous system, an increase in dopamine synthesis could cause a higher turnover of this amine, resulting in abnormal quantities of dopamine available at central or sympathetic receptors. It would also explain the rise in excreted HVA. There is still no causal evidence to indicate a malfunction in catecholamine synthesis in dysautonomia, but speculation about such a relationship is quite valid.

4.2. METABOLISM IN PSYCHOSES

4.2.1. DIMPEA Theory

The provocative study by Friedhoff and Van Winkle (1963) of a group of schizophrenics described a urinary substance, later identified as 3,4-dimethoxyphenylethylamine (DIMPEA), which was present, in 15 of 19 schizophrenics and absent in all 14 normal controls. Chemically, DIMPEA bears a structural resemblance to 4-hydroxy-3-methoxyphenylethylamine, a normal metabolite of dopamine. Friedhoff and Van Winkle have pointed out that even though dopamine is normally methylated in the 3- or meta-position, resulting in 4-hydroxy-3-methoxyphenylethylamine, there can also occur para-O-methylation in the 4-position. Individual studies have shown that dopamine may be converted to 3,4-dimethoxylated derivatives. In addition, 3,4-dimethoxyphenylacetic acid, an acid metabolite of DIMPEA, has been isolated from urine of schizophrenics after administration of dopamine-7H^3. Also reported was evidence that DIMPEA can be converted to 3,4-dimethoxyphenylacetic acid when administered to schizophrenic patients. Similarly, dopamine was converted to 3,4-dimethoxyphenylacetic acid using liver obtained by biopsy from schizophrenics as a source of the enzyme. The obvious question is whether schizophrenics or even some schizophrenics are able through an altered enzyme system to diorthomethylate a basic catecholamine into a substance not apparently characteristic of normal urine (Fig. 39). The finding is still being debated by some who confirmed this phenomenon and others who did not. The most recent confirmation has come from Bourdillon and Ridges (1967), who reported on a fairly extensive and well-controlled group. Disturbed methylation in schizophrenics was previously suggested by Harley-Mason (1952), who pointed out that catecholamine methylation could occur in one or both of the hydroxy groups in the 3- and 4-positions. Although DIMPEA

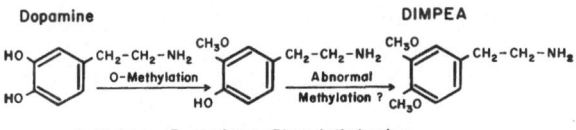

Fig. 39. Possible abnormal reaction of metabolized
dopamine leading to production of DIMPEA.

does not produce any psychopathic effect upon administration to humans, it is known to induce a catatonic-like reaction in animals.

Other investigations have revealed that ingestion of the methyl donor methionine by schizophrenic patients appears to exacerbate their symptoms. Inversely, the methyl acceptors niacin and niacinamide have been reported as having beneficial effects when they were given to a few schizophrenics. This drug treatment, however, has not been confirmed by study with a well-controlled group of schizophrenics. The fact that injection of DIMPEA in animals is able to cause some rigidity and akinesia has been correlated with the catatonia of schizophrenia and the bradykinesia of parkinsonism and would indicate that this compound may affect basal ganglia structures. DIMPEA or the so-called "pink spot" has been reviewed by Barbeau (1967), who concluded after chromatography of urine of patients with schizophrenia and parkinsonism that this substance is more often present, in larger amounts, than in normal subjects. Perry *et al.* (1967) reported a "pink spot" consisting of three noncyclic compounds found in the urine of schizophrenic patients. Two of these compounds were shown to be monoacetylcadaverine and monopropionylcadaverine. The three compounds that yielded the "pink spot" similar to that of DIMPEA could not be separated by paper chromatography using any of the solvent systems previously used by investigators who had reported "pink spots." Since certain phenothiazine metabolites such as norchlorpromazine sulfoxide also yield a "pink spot," the possibility exists that the "pink spots" reported to date in schizophrenic or parkinsonian patients may represent amines formed by intestinal bacteria or metabolites of these phenothiazines.

The disturbed methylation in schizophrenia was first proposed by Osmond and Smythies (1952) based on unpublished observations by Harley-Mason, who hypothesized that the biogenesis of adrenaline is a trans-methylation of noradrenaline, the methyl group arising from the methionine or choline. Noteboom (see Osmond and Smythies, 1952) had earlier found that in a series of phenylethylamines tested on animals the most potent in inducing catatonia was DIMPEA, a substance with a structure identical to mescaline except for its lack of a hydroxyl group on the side chain. Osmond and Smythies then suggested that schizophrenia could result from a specific disorder of the adrenals wherein a metabolic transformation disorder results in the production of a mescaline-like compound. Mescaline in the system would, of course, produce the sensory, motor, behavioral, thought, and mood disorders frequently found in schizophrenia. It will remain to be seen whether this theory has any worth at all in clarifying psychotic disorders.

4.2.2. Serotonin Hypothesis

In 1954, working independently, Woolley and Shaw in the United States and Gaddum (1953) in England suggested that a faulty mechanism in the metabolism of serotonin could be the basis of psychotic behavior. Woolley (1962) later extended this idea into a serotonin hypothesis, in which he pictured schizophrenia and other mental diseases as arising from the abnormal metabolism of serotonin in the brain. In support of this postulate, Woolley has pointed out the action of LSD in preventing or abolishing smooth muscle contraction caused by serotonin. Additional support for the serotonin hypothesis was based on more empirical evidence and can be summarized as follows: (1) interference of tranquilizing drugs with the functioning of serotonin (for example, chlorpromazine and reserpine are known to prevent the normal action of serotonin on tissues and the decrease in serotonin content of the brain, respectively); (2) increase in the serotonin content of the brain by various drugs which relieve the symptoms of depression; (3) abnormal excretion of the serotonin metabolite 5-hydroxyindoleacetic acid in the urine of schizophrenic patients; (4) pharmacological increase of serotonin by various drugs which appear to exacerbate the symptoms of patients with schizophrenia; (5) structural and functional similarity between lysergic acid diethylamide (LSD-25) and serotonin analogues.

The first observation, that drugs such as reserpine and chlorpromazine affect the metabolism of serotonin in tissue and brain, is well documented. The action of reserpine on the monoamines is well known, the effect being termed "the reserpine syndrome." Reserpine in low dose drastically reduces the brain levels of the monoamines without modifying the rate of their synthesis. This decrease in the levels of monoamines in the central nervous system persists even after the disappearance of reserpine from the brain. Reserpine probably inhibits catecholamine uptake into storage sites, and the same effect holds true for serotonin. By blocking the uptake of monoamines into storage sites, reserpine effectively depletes these amines from the brain. In addition, reserpine also releases serotonin from all tissues containing it. This is seen particularly in blood platelets and the gastrointestinal tract. It is not yet settled whether the reserpine syndrome occurs as a result of mobilization of serotonin or of noradrenaline or of both. Brodie et al. (1966a) believe that serotonin is the mediator in the reserpine effect. Brodie suggests that the reactions following reserpine administration and involving the central and peripheral parasympathetic systems classify the amine as belonging to the trophotropic system postulated by Hess (1936). He argues that the stimulating effects seen after monoamine oxidase inhibitors and reserpine are a result of paradoxical stimulation of central synapses occurring through the release of large quantities of brain serotonin.

However, if serotonin is pharmacologically diminished in the brain by the use of p-chlorophenylalanine, no sedation occurs. This has been well documented by Koe and Weissman (1966b) and has been confirmed repeatedly by other investigators. The overall picture of animal and human studies on reserpine indicates a closer correlation to catecholamine depletion in specific brain regions. In addition, it should be noted that the administration of the serotonin precursor 5-HTP does not modify or potentiate the action of reserpine in animals, while the reverse is true after the dopamine precursor L-dopa is administered. By contrast, chlorpromazine or the so-called major tranquilizers produce a variety of reactions very similar to those of reserpine but without changing the amine levels or the storage mechanisms. The spectrum of reactions produced by chlorpromazine treatment would undoubtedly fill many pages dealing with psychomotor disturbances. Some of the symptoms include violent coarse trembling, stiffness of muscles, motor restlessness (akathisia), inability to remain seated (dysakinesia), as well as many of the classical symptoms observed in Parkinson's disease such as akinesia or poverty of movements, masklike expression, oculogyric crises, twisting movements of the extremities, and, in more severe cases, opisthotonus. These symptoms can occur very early in the treatment or they may appear very late. Although the signs may be alarming, they are usually successfully reversed by the use of antiparkinsonism agents, among which is L-dopa. Interestingly enough, chlorpromazine can antagonize the psychotomimetic actions of LSD in humans. However, more work is needed to substantiate a specific action of this drug on brain serotonin.

The second observation, that antidepressants such as the monoamine oxidase inhibitors affect brain serotonin concentrations, is also nonspecific. Such MAO inhibitors as iproniazid and nialamide can and do increase locomotor activity, as seen by animal studies. However, this activity seems to be more directly related to increase in noradrenaline than to increase in serotonin. Our own studies using nialamide in rats suggest an increase in monoamines located at synaptic nerve terminals and neuronal cell bodies, as shown by the histochemical fluorescence technique. The increase in monoamines is equally evident in noradrenergic as well as serotonergic nerve terminals and cell bodies. A slight tranquilizing effect is observed after 5-HTP administration in rats previously treated with a dopa decarboxylase inhibitor to facilitate penetration of the amine precursor through the blood–brain barrier. The opposite effect is seen after L-dopa administration, that is, increased locomotor activity and restlessness. After prolonged treatment with the MAO inhibitors, there is a decrease in the rate of excretion of the catecholamine metabolites 3-methoxy-4 hydroxymandelic acid and homovanillic acid as well as the serotonin metabolite 5-hydroxyindoleacetic acid. This lowered excretion rate in the urine of animals and man reflects an increase in the endogenous brain concentration of these monoamines. This increase can be linked to the inhibition of MAO located at the mitochondrial level in nerve terminals, and while MAO inhibitors do not alter the turnover rate for the monoamines they do prolong the half-life of injected tyramine and tryptamine.

The third observation, that of increased excretion of 5-hydroxy-indoleacetic acid in the urine of schizophrenic and psychotic patients, is more difficult to explain. Constantinidis (1965) reported that the average daily urinary excretion of 5-hydroxyindoleacetic acid of a group of typical psychotic patients was significantly lower than that of a control group and much lower than that of either depressed patients displaying motor inhibition or typically manic patients. Their study led them to conclude that there is a direct correlation between the level of excretion of 5-hydroxyindole-acetic acid and psychotic syndrome. Further studies on the rate of excretion of serotonin metabolites in the urine of schizophrenics failed to add any positive knowledge to our understanding of these disorders. Verster (1963) reported successful treatment of four patients with acute mania by using the antiserotonin substance methysergide. Recently, Dewhurst (1968) confirmed this by reporting a dramatic improvement in manic patients by use of the same compound. The theoretical basis for this form of treatment is that the metabolism of indoles, particularly serotonin, is disturbed in these disorders. Haškovec and Souček (1968), motivated by the success of the two previous studies, set up an acute clinical trial of methysergide for 8–14 days, giving a maximum dose of 6 mg per day and substituting a placebo for 2–6 days thereafter. Their results appear to confirm the studies by Verster and Dewhurst. Fieve et al. (1969), attempting to confirm the favorable effects of methysergide in manic patients, repeated this experiment but without success. The same authors reported a favorable result using lithium to replace methysergide as the control drug. The results of all these studies indicate that further work is necessary before a tentative proposal can be accepted.

The fourth observation, that increasing 5-HT in the brain by pharmacological means results in alleviation of depression, is, at the present time, unconfirmed. Lithium, for example, which lowers the activity of serotonin nerve cells or inhibits the impulse-stimulated release of serotonin at the nerve terminals, has been used successfully in many human trials for the treatment of manic patients. The serotonin precursor 5-hydroxytryptophan has not been tried to the best of our knowledge in human clinical trials for acute depression. However, since this serotonin precursor has been shown to penetrate the blood–brain barrier in only very small amounts, it seems logical that if this form of treatment were adopted, pretreatment with a peripheral dopa decarboxylase inhibitor would be necessary.

The final observation, on the structural and functional similarity between LSD and serotonin derivatives, is still a subject of controversy. In looking at the formula for LSD (Fig. 40), it is evident that this substance possesses a 4-substituted tryptamine element as part of its molecular structure. Other psychotomimetic compounds which are simple tryptamine derivatives are psylocin, which has been identified as 4-hydroxy-N,N-dimethyltryptamine, and psilocybine, which is its phosphoric acid derivative. Cerletti *et al.* (1968) found a close relationship between the 4-substituted dimethyltryptamines and LSD in the ability of these compounds to elicit an antagonistic effect on serotonin-stimulating pyrogenic activity and to increase the knee-jerk reflex in cats. These findings suggest that 4-hydroxylation of dimethyltryptamine compounds increases their potency in the living system and alters their effect on behavioral reactions. Moreover, the hallucinogenic properties of some 4-substituted N,N-dimethyltryptamines (in particular, 5-methoxylated dimethyltryptamine derivatives) and their interaction with serotonin in metabolism suggest a similarity of action to LSD. In addition, N-alkylated derivatives of tryptamine such as dimethyltryptamine have been found to be hallucinogenic in man. These compounds have been shown by Szara (1964) to be metabolized through a 6-hydroxylation dealkylation and deamination to form corresponding intermediates. A recent report by Heller *et al.* (1970) describes the presence of dimethyltryptamine in the blood of five patients with acute schizophrenia. The dimethylated indoleamines were not detected in a group of nine chronic schizophrenics, in two normal subjects, or in a depressed subject. Szara (1964, 1968) has suggested that 6-hydroxylation of indolealkylamines (see Fig. 41) might be necessary for the hallucinogenic reaction but not sufficient to account for the psychotropic action of this drug. On the other hand, LSD and 5-HTP can produce similar functional effects in the rat's spinal cord and brain, which, according to Andén *et al.* (1968), indicate essential stimulating action by LSD or serotonin receptors. These effects were not observed when the brominated isomers of LSD and methysergide were used.

This same group, using the histochemical fluorescence technique and biochemical assays, also reported that LSD reduced the turnover rate of brain and spinal cord serotonin, while the turnover of brain noradrenaline,

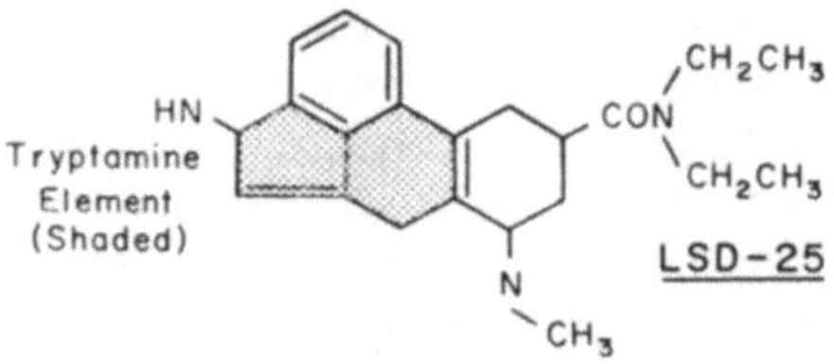

Fig. 40

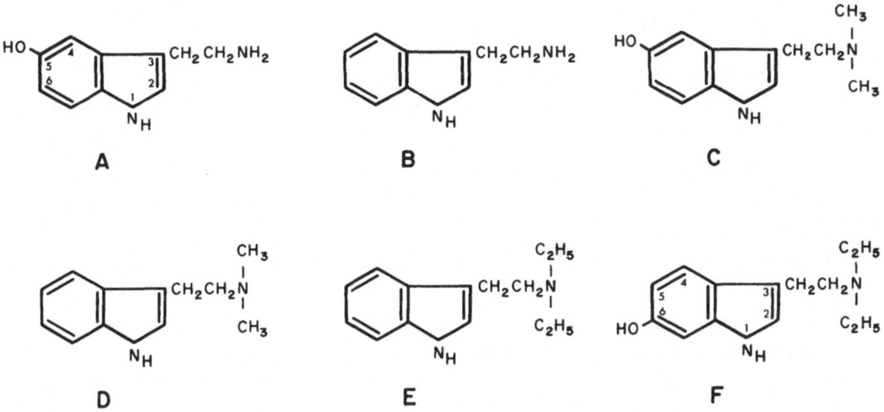

Fig. 41. Szara (1964, 1968) has proposed that 6-hydroxylated tryptamine derivatives, particularly dimethyltryptamine (DMT) and diethyltryptamine (DET) (see D and E), are important as precursors or metabolites of psychoactive substances. His evidence is that (a) synthetic 6-fluoro-DET has no psychotomimetic action in man; (b) liver microsomes can 6-hydroxylate DMT and DET (both psychotomimetic), while dibutyl and dihexyltryptamines which are not 6-hydroxylated are inactive; (c) 6-hydroxy-DET (F) can affect behavior patterns in rats, mice, and monkeys, but its nonhydroxylated isomer cannot; and (d) psychotomimetic severity in man after DET is proportional to the amount excreted in urine of the 6-hydroxy metabolite.

but not dopamine, was somewhat accelerated. They concluded that the slowdown of serotonin turnover by LSD could be due to the negative feedback mechanisms produced by direct stimulation of central serotonin receptors. Woolley's (1962) report on the tetanic contraction of human oligodendroglia *in vitro* caused by serotonin is not without interest in view of the work by Elvidge and Reed (1938) linking histopathological state of these cells with schizophrenia and manic depressive psychosis. This line of experimentation, however, has never been further pursued. The argument by some workers that the brominated isomer of LSD, BOL-148, can exert antagonistic effects on serotonin while lacking hallucinogenic properties would seem to be unfair, since the central action of BOL-148 has never been determined and its blood–brain barrier permeability is not known. Besides that, in the field of neuropharmacology two analogues may have similar actions on a physiological system while behaving differently at central receptor sites. This is not to say that the serotonin hypothesis does not have some grave contradictions; rather, it is an attempt to judge the local merits of such a proposition while remembering that nonconfirmatory evidence does not necessarily obviate positive findings.

4.2.3. Catecholamine Hypothesis

A review of psychoaffective states involving monoamines would not be complete without a discussion of the roles of dopamine, noradrenaline, and adrenaline in these states. Evidence linking catecholamine metabolic changes with alterations in the affective state is fairly recent. The first was obtained by Vogt (1954) in her classic experiment detailing the localization of these catecholamines in the animal brain. The fact that the catecholamines are concentrated in central areas representing the sympathetic system is suggestive of their importance in the functioning of these brain regions. These regions include the central gray matter of the mesencephalon, the reticular formation, the hypothalamus, the septum, and the brain structures directly related to the so-called limbic system. However, even before the localization of catecholamines in the brain, Marañon (1924) had experimented with adrenaline in human subjects. He infused adrenaline into volunteers, who reported symptoms of fear, anger, anxiety, and well-defined emotional states. The patients found these to be "cold" emotions, not real feelings but rather subjective states of being. Marañon called them "as if" reactions, since the patients did not feel anger or anxiety but felt "as if" they were angry or anxious. Since the penetration of adrenaline is very poor through the blood–brain barrier, one could propose that the effect of this substance is felt either in peripheral receptors or through microquantities that are able to penetrate into the brain substance, such as in the hypothalamus, which has been shown to be more permeable to sympathetic amine penetration than other brain regions.

Adrenaline has also been implicated as a possible causal factor in schizophrenia. It has been shown that serum from schizophrenics can abolish or reverse the usual pressor response reactions observed after topical application of adrenaline to exposed cerebral cortex of rabbits. Sera from normal individuals, pregnant women, or persons with various organic diseases did not have any effect on the adrenaline hypertensive reaction. These data support the evidence of Heath and Krupp (1967) that a substance isolated from serum of schizophrenics is capable of causing schizophrenic symptoms in nonschizophrenic subjects and marked encephalographic changes when administered to monkeys. Further studies with rabbits, however, have failed to confirm any changes in the concentration of adrenaline, noradrenaline, and dopamine in the hypothalamus, cerebral hemispheres, or brain stem after injection of serum from schizophrenics. Also investigating this effect, Losovski (1965) found that while serum of patients afflicted with nuclear forms of schizophrenia did not cause any significant changes in the noradrenaline content of the hypothalamus of rabbits, sera from periodic schizophrenic patients caused a substantial increase in hypothalamic noradrenaline levels. A relationship between the oxide derivatives of noradrenaline has been suggested (Hoffer and Osmond, 1959) as an etiological factor in schizophrenia. Adrenochrome and adrenolutin were alleged to cause mental changes characteristic of schizophrenia when injected into normal human subjects. The basis given for this was that in schizophrenia a high phenolase activity degrades adrenaline to some extent by increasing sympathetic activity. The result is a toxic quinone resulting in adrenochrome and adrenolutin. Recent studies attempting to confirm this finding have in fact found no basis for any of these assumptions. Subsequent reports by these same authors (Hoffer *et al.*, 1957) claimed miraculous recovery of schizophrenics after nicotinamide or nicotinic acid therapy. Not only were these claims not confirmed by other workers, but in a study by Meltzer *et al.* (1969) nicotinic acid was found to exacerbate the symptoms of some schizophrenics. Catecholamine excretion products such as normetadrenaline, metadrenaline, and vanillylmandelic acid were reported by Gjessing (1964) to increase psychotic episodes in patients with periodic catatonia and to return to normal levels when the psychotic attack subsided. Proponents of the catecholamine hypothesis have further divided their views in proposing that either dopamine or noradrenaline is more important in psychoaffective disorders.

In the brain, only enzymatic hydroxylation separates dopamine from noradrenaline concentrations, and drugs which either decrease or increase the levels of these two catecholamines can produce alterations in gross behavior patterns. It is clear that before any claims can be made as to the causal relationship between either catecholamine and affective disorders one must make sure that only that amine is affected and that the brain concentration of the other amines has not been altered. To do this, one can use drugs' which specifically block the action of certain monoamines. These enzyme blockers and inhibitors were discussed in Section 3.4. One of these drugs, disulfiram, which selectively lowers the concentration of noradrenaline in the brain by inhibiting dopamine-β-hydroxylase, is contraindicated in the presence of psychosis. How the specific decrease of noradrenaline in the brain is exacerbated in psychotic individuals by the use of selective inhibitors is not yet known. Reis and Fuxe (1969) reported that there appears to be a direct relationship between the magnitude of sham rage produced by brain stem transection in the cat and the decrease in brain stem noradrenaline. The attacks of rage in these animals can be increased by protryptiline and inhibited by haloperidol. These authors concluded that the release of noradrenaline by brain stem neurons appears to be essential for appearance of this behavior. In a follow-up report, Reis *et al.* (1970) reported a high degree of correlation between the level of noradrenaline in the brain stem and the magnitude of excitement individual cats receiving L-dopa, phenypromazine, reserpine, or disulfiram, alone or in combination. These authors believe that the excitement was probably due to the release of newly synthesized noradrenaline.

Another drug used for manic disorders is lithium carbonate, a compound recently added to the neuropsychopharmacological armamentarium. Messiha *et al.* (1970) reported that following lithium carbonate treatment of a group of manic patients, the excretion of dopamine in urine, which had been significantly elevated in six of seven patients, returned to normal. These workers suggested that increased excretion of dopamine is one of the biochemical changes associated with a manic phase of an affective disorder. They further proposed that lithium ions in low concentrations inhibit the electrical stimulation–induced release of noradrenaline and serotonin. Their study was based on an observation by Katz *et al.* (1968) that following treatment with lithium carbonate, the excretion of dopamine in manic patients decreased to normal levels.

Other investigations suggest that in acute depression there is a decrease in dopamine concentration in the brain as reflected by urinary excretion rates of dopamine metabolites. Attempting a more direct evaluation of the hypothesis that depression is associated with a functional decrease in brain catecholamines, Goodwin and Brodie (1970) reported their results with ten patients suffering from depression treated with a dopa decarboxylase inhibitor (MK-485) followed by L-dopa. Of this group, only four patients improved on the combination, three of these relapsing when a placebo was substituted, while three patients showed no response and three were apparently made worse by the treatment.

The fact that an increase in dopamine level in the brain of these patients helped some of them suggests that a more precise method of classification of psychotic disorders, based perhaps on biochemical analyses, would be worthwhile. This does not rule out the possibility that to increase the dopamine level in this condition is completely worthless, or for that matter does not mean that idiosyncratic behavior cannot be globally classified under the major heading of depression. It must be kept in mind that these studies are only in the preliminary stages and that we are merely scratching the surface of a complex problem that may have other etiological considerations. It is clear that human studies are somewhat limited in that only urine, cerebrospinal fluid, and blood testing, and in extreme cases small tissue biopsies, can be considered in view of the patient's safety. The limitations of animal studies are also obvious. Psychoactive drugs injected into rats will not tell us anything about their mental state, and even though amine changes may be observable by various histochemical and biochemical techniques, our extrapolation of these results to human behavior will always be speculative and liable to pious but well-founded criticism.

4.2.4. Various Affective States and Stress

An affective state is one that involves mood changes such as anxiety, fear, or anger. These mood changes can be normal or abnormal in nature. The ability of monoamine oxidase inhibitors to clinically alleviate depressive moods in patients and the effect of reserpine to produce a state of endogenous depression in human subjects suggest that central monoamines have a role in such affective states. In view of this, many studies have been devised to relate the release of noradrenaline and adrenaline to emotional behavior. One early report by Funkenstein *et al.* (1954) on both normal and psychotic subjects tested the cardiovascular response to stress. Blood changes were noted in aggressive or angry states similar to the reaction noted after noradrenaline infusion. Other cardiovascular responses were seen in anxious or depressive periods, this time mimicking the response observed after adrenaline infusion. Various forms of stress can also produce alterations of central monoamine levels in animals and man, as previously discussed. Barchas and Freedman (1963) put rats through a series of physical and chemical stress situations and found moderate increases in the brain serotonin levels with concomitant noradrenaline decrease in such conditions as swimming, wheel running, exposure to cold, and LSD administration. No change in monoamines was found after electroshock treatment to feet, nor after 72 hr of food and water deprivation following adrenalectomy, nor after anoxia for 10 min in a nitrogen chamber. Loss of endogenous noradrenaline after stress would indicate a synaptic release of this amine and would correlate with anxiety situations. Von Euler and Lundberg (1954) have also shown an increased excretion of both noradrenaline and adrenaline in aircraft pilots after moderately stressful flights. Since adrenaline is excreted from the adrenal medulla after hypoglycemic stress and cold, several investigations have been carried out on adrenalectomized animals in order to study this effect on cerebral amines. One such study reported that adrenalectomy can facilitate the action of reserpine and in addition increase the brain serotonin concentration. It was found that following adrenalectomy, electroshock treatment did not alter noradrenaline in animals, even in control animals not operated on. The MAO inhibitor nialamide does not seem to affect brain serotonin after adrenalectomy. Angel and Burkett (1966) reported on the importance of the adrenal gland in relation to the permeability of the blood–brain barrier. After bilateral adrenalectomy, the blood–brain barrier became more susceptible to penetration by cocaine, but after unilateral adrenalectomy, bilateral oophorectomy, or sham operation, no changes in the brain barrier permeability were evident in the animals. Mild conditions of stress, such as mental stress induced by psychological tasks, appear to increase adrenaline excretion as compared to that during intervals of testing inactivity. It would appear from these studies that there is a relationship between catecholamines and certain mood situations. Thus noradrenaline excretion is relatively increased among aggressive or angry subjects, while adrenaline excretion preferentially accompanies reactions involving anxiety or mild stress conditions. Since adrenaline and noradrenaline do not cross the blood–brain barrier except perhaps in minute amounts at the level of the hypothalamus, it is possible that peripheral release or administration of these amines can produce affective changes in a subject by direct action on hypothalamic receptors. Elmadjian *et al.* (1957) found increases in noradrenaline secretion accompanied by lesser increases in adrenaline excretion in hockey players during active competition and also in patients with mental disorders during aggressive emotional outbursts. The amine excretion was less pronounced in hockey players not actively participating in the game than in patients, for example, attending psychiatric staff conferences.

Visual images can also affect monoamine excretion rates. Scenic films have been associated with a reduction in urinary noradrenaline and adrenaline excretion, while emotionally provoking films such as comedies or tragedies increase the excretion of both amines. Brain catecholamines have been shown by Häggendahl (1967) to be decreased after exposure of rats to high pressure in the presence of oxygen. Convulsions ensuing after exposure of rats to oxygen with carbon dioxide added may affect or disrupt the balance of catecholamine-containing neurons in the brain. Asphyxia is known to lower the amount of noradrenaline in the brain provided that it causes medullary secretion and severe hypoxic hypoxia—which can diminish brain catecholamine levels in cats, with specific noradrenaline reduction occurring in the hypothalamus. Karyometric changes in the hypothalamus following stress in rats have been reported by Mitro (1965). When rats were immobilized for $2\frac{1}{2}$ hr each day for 7–42 days, cellular changes in the dorsomedial and arcuate nuclei were observed after a few days. After several weeks, the rats adapted to the stressful condition, and only repeated stress resulted in cell alterations in the ventromedial nucleus. This finding is suggestive of an important relationship between the hypothalamic ventromedial region and adaptation to repeated stress.

Defense reaction or sham rage is elicited in the cat by electrical stimulation of the amygdala or hypothalamus and is accompanied by a lowering of the noradrenaline concentration in the brain as well as of noradrenaline and adrenaline in the adrenals, while cerebral serotonin and dopamine remain unchanged. Fluorescent histochemical studies have demonstrated that noradrenaline depletion in the brain is the result of a *mean* outflow within the axon terminals belonging to the noradrenaline-containing nerve cells. Since there do not appear to be appreciable changes in the serotonin content during rage, it is probable that this amine plays little or no role in this behavioral reaction. By contrast, noradrenaline may be implicated in the mechanism concerned with certain behavioral states including excitation, escape, and defense reactions or the so-called fright–flight–fight responses. Studies using radioactive noradrenaline injected into animals subjected to various types of stress show that the amine is more rapidly depleted from the tissues that in nonstressed control animals. The accelerated turnover of the radioactive noradrenaline in the stressed animals may be an indication of an increased impulse flow from the sympathetic nerves produced by the different stress factors. Also, the increase in radioactive noradrenaline turnover is independent of vascular responses measured as vasoconstriction or vasodilatation, although vasomotor response is usually seen after α- and β-blocking adrenergic agents. Electroconvulsive shock in animals has also been found to result in the increased turnover of noradrenaline and its subsequent decrease in the brain. It is not yet possible to conclude from these studies that the released noradrenaline reaches certain receptor sites in the central nervous system to stimulate them, after which the released noradrenaline could be inactivated into its methoxy derivatives by catechol-*O*-methyltransferase (COMT). But in certain forms of stress such as electroconvulsive shock, there is a general increase in the methoxy and dihydroxy compounds of noradrenaline. This suggests the possibility that noradrenaline released after shock treatment may be metabolized by MAO inside the cell and by COMT at the receptor site, thus inactivating completely its mechanism of action.

The noradrenaline precursor L-dopa, given to cats treated with a MAO inhibitor, can produce a state of excitement resembling sham rage. This condition results in a very alert, very irritable animal, which is easily startled by noise and whose main preoccupation seems to be to escape from the handler. These typical drug responses to L-dopa have been reported by Reis *et al.* (1970), who also point out that the drug disulfiram, which inhibits the formation of noradrenaline in the brain, is capable of minimizing or abolishing the sham rage behavior produced by the L-dopa and the MAO inhibitor. It has been known for a long time that electrolytic lesions in the septal region of the rat brain can produce stereotypical behavior and a very severe rage reaction in which the animal will not only attack its cagemates but also anything that is put in its path. The animal is extremely irritable, jumping up in the air several feet when its back is stroked and showing autonomic manifestations such as piloerection, salivation, and motor restlessness. It should be pointed out that the septal region contains a great amount of dopamine as well as noradrenaline and that destruction of this region may affect neurotransmission in such areas as the hypothalamus, which can regulate emotion and affective behavior. Some authors have described a rage reaction in rats occurring after administration of a MAO inhibitor plus L-dopa. The animals thus treated are described as "enraged." Vocalizing and running are also part of their stereotypical behavior pattern. Our own experience, however, is that the drug treatment of rats with a MAO inhibitor plus L-dopa and the electrolytic lesion of the septum and other brain regions produce entirely different reactions. Besides severe autonomic effects observed with the drug treatment but hardly noticeable with the septal lesion, there is also an important difference in the pattern of excitability. This can best be described as irritability in the drug-treated animal, and rage, with all its psychic manifestations, in the rat with the electrolytic septal damage.

It would seem then that dopamine and noradrenaline in the brain play an important role in the normal response to stress and that drugs which modify the metabolism of these amines are also capable of altering the affective state. These behavioral reactions evoked by either stress or drugs are manifested in both man and animal, with some variability in the degree of response. It need hardly be pointed out that one of the most important of all psychiatric tools for the treatment of acute or chronic psychosis is convulsive therapy. There are two types: convulsions induced by drugs such as hexafluroethyl ether (Indoklon) and convulsions produced by electrical shock applied through electrodes on the scalp. It has been seen before that such stress in both animals and man can mobilize monoamines in the brain. However, it is still too early to make any assumptions concerning monoamine manipulation as a form of treatment in patients with neuropsychiatric disorders. Several well-controlled studies of humans by use of these parameters would undoubtedly clarify the role of monoamines in specific disorders.

Within the last few years, several theories have evolved concerning the role of noradrenaline or dopamine in psychotic disorders. The so-called noradrenaline hypothesis with respect to depression and other intense mood changes in man has been suggested by various investigators. Although the evidence accumulated thus far is only indirect, it is nevertheless of interest to review it in view of its important implications. Schildkraut and Kety (1967) have advanced the notion that noradrenaline and serotonin work in reciprocal relationship with each other as far as depression and mania are concerned. They therefore speculate that norepinephrine is responsible for such mood changes as pain, elation, pleasure, and depression. Thus, high concentrations of noradrenaline in the brain would be causally related to mania and low noradrenaline concentrations would cause depression. Kety (1970) further speculates that serotonin is the all-important factor involved in mood changes and that changes in the concentration of this indoleamine in the brain determine whether or not such mood changes will result in depression or mania.

This, of course, is a simplification of a very complex problem that may involve other substances besides noradrenaline and serotonin—which is readily admitted by the proponents of the hypothesis. The first line of evidence to support the noradrenaline hypothesis comes from the neuro-pharmacological effects of certain drugs on the brain levels of noradrenaline. As such, the current antidepressant drugs are capable of increasing the concentration of free noradrenaline at the site of central adrenergic receptors. The MAO inhibitors achieve a similar effect by preventing the degradation of noradrenaline located intraneuronally. Noradrenaline may also be potentiated by the use of tricyclic antidepressive agents, which probably act to reduce the permeability of the cell membrane to noradrenaline, thus impairing noradrenergic uptake at the cellular level while providing a higher concentration at the receptor sites. One fact which lends support to the noradrenaline hypothesis is the ability of reserpine to deplete noradrenaline from its storage sites, although not selectively, and to produce in patients a state clinically indistinguishable from endogenous depression. By the same token, α-methyldopa, which inhibits the transformation of L-dopa into dopamine by inhibition of dopa decarboxylase endogenously, can also reduce the brain concentration of active noradrenaline and at the same time provoke a typical depressive mood in humans. Another drug which increases the availability of noradrenaline at central synapses is amphetamine, which can induce excitement in both animals and man, and its effect can be reversed by pretreatment with α-methyltyrosine, an inhibitor of catecholamine synthesis. Moreover, the effects of amphetamine upon the rate of self-stimulation in rats can also be blocked with reserpine, which may act to bring the noradrenaline in the brain back to its original level. Studies by Clouet and Ratner (1970) show that morphine and other narcotic analgesic drugs may interact with central monoamines. These two investigators suggest that in animals the acute injection of morphine causes a release of dopamine and noradrenaline from the brain. Thus, the depressive action of morphine may be related to noradrenaline release or to changes in the rate of turnover of the two amines in the brain. Similarly, cocaine, which produces a euphoric effect in man, has been found by Dengler *et al.* (1961) to increase the endogenous levels of noradrenaline following intraventricular injections of noradrenaline-H^3, which, according to Glowinski and Axelrod (1966) may allow the penetration of the radioactive noradrenaline into the brain by avoiding the blood–brain barrier. A summary of the effects cn mood of these drugs as well as their presumed effects on noradrenergic receptors is given on Table 15.

Besides the bulk of the pharmacological evidence that exists for drugs that affect the noradrenaline concentrations in the brain and as a result affect mood, additional evidence of a physical nature is also available. Studies by Kety (1970) show that after electroconvulsive therapy in animals, a release of brain noradrenaline occurs. As electric shocks are repeated, a greater stimulation of the production of noradrenaline results in a 30%–50% increase in brain levels. This form of treatment is used by thera-

Table 15 [a]

Drug	Effect on mood (humans)	Effect on Na (animals)	Effect on NA receptors
Monoamine oxidase inhibitors	Antidepressant	Inhibits deamination, $>$ NA, $>$ NMN, $<$ DCM	Increase
Imipramine Desmethylimipramine	Antidepressant	Inhibits cellular uptake of NA, $<$ turnover NA, $>$ NMN	Increase
Cocaine Amphetamine	Stimulant	Inhibits cellular uptake of NA, $>$ NMN, $<$ DCM, discharge of NA (amphetamine)	Increase
ECT	Counteracts depression	Increases neuronal discharge of NA, $>$ NMN	Increase
Reserpine	Depression	Depletes NA	Decrease
Lithium salts	Counteracts mania	$<$ NMN, $>$ DCM	Decrease

[a] Key: increase, $>$; decrease, $<$; normetadrenaline, NMN; deaminated catechols, DCM; electroconvulsive shock, ECT.

pists for severe depressions in humans and often results in dramatic improvement. The exact reasons for this improvement are nevertheless not known, but the finding of increased noradrenaline in the brain is compatible with the mood elevations seen after electroconvulsive therapy. We have discussed the use of lithium carbonate in the treatment of mania and the animal studies that indicate that lithium increases the intraneuronal breakdown of noradrenaline and thus decreases the noradrenaline available at central adrenergic receptors. Physiological stress may also be related to noradrenaline levels in the brain. Barchas and Freedman (1963) and Maynert and Levi (1964) have shown that life-threatening stress, as well as pain and fatigue, is associated with a reduction in brain noradrenaline.

Recent evidence by Cohn *et al.* (1970) concerning the role of COMT in depressed women appears to strengthen the noradrenaline hypothesis. This group found a significant reduction of COMT activity in red blood cells of women suffering from primary affective disorder (depression). The COMT activity in the blood of depressed men and in that of normal men and women was unchanged. COMT was also unmodified in schizophrenics of both sexes. The deficiency of catecholamines may result in (1) reduction of catecholamine release from nerve endings and adrenal glands and (2) decrease in response at the adrenergic receptor sites. The fact that depressed women respond more favorably to antidepressant drugs than men indicates some association between specific sex hormones and amine mobilization.

A study by Huttunen (1971) on the effects of stress in pregnant rats revealed for the first time persistent changes in the rate of turnover of brain noradrenaline in the offspring of these rats. If these results are confirmed, it would provide a very attractive hypothesis not only in support of noradrenergic brain changes after stress but also on the genetic implications of such stress on pregnant animals. It is at present extremely difficult to obtain much evidence on the rate of turnover and changes in the metabolic concentrations of brain noradrenaline in humans, since one cannot use direct methods of tissue analysis on patients suffering from severe depression and apply these findings in a meaningful statistical analysis by comparing them to values for normal or nonpsychotic individuals. The merits of the noradrenaline hypothesis lie in the fact that it can provide a framework of reference for the use of current antidepressant drugs and an approach toward further research on the central role of noradrenaline in affective disorders.

This takes us to a second hypothesis recently advanced by Stein and Wise (1971), which for lack of a better name we shall call the dopamine hypothesis of schizophrenic reactions. If the lowering of noradrenaline produces depression and its increase results in manic symptoms, then the concentration of endogenous dopamine in the brain also produces enhancement of motor excitation in experimental animals. The dopamine hypothesis attempts to explain the action of 6-hydroxydopamine (6-OHDA) on the central nervous system after its administration in animals. Tranzer and Thoenen (1968) reported that 6-OHDA selectively destroyed peripheral adrenergic nerve endings in various species. After intraventricular application of 6-OHDA, the noradrenaline concentration in various peripheral tissues was also decreased, this lasting for several weeks. Bartholini *et al.* (1970) studied the effects of intracerebral administration of 6-OHDA using both histochemical fluorescence and electron microscopic examination. They reported that intracerebral injections of 6-OHDA had relatively specific action on catecholaminergic neurons. Two days after a single 6-OHDA treatment, the brain content of dopamine and noradrenaline was reduced to 68% and 35% of control levels, respectively, while serotonin only decreased to 88% of the control levels. Histochemical fluorescence revealed the disappearance of catecholamine fluorescent neurons in brain after the intracerebral administration of 6-OHDA. Ultrastructural examination of the periventricular hypothalamic and caudate nuclei revealed no changes compared to saline-treated controls. However, after two injections of 6-OHDA, ultrastructural changes were found in both areas, although no changes in the levels of noradrenaline were seen. Constantinidis *et al.* (1971) confirmed these results using 200 μg of 6-OHDA intraventricularly administered. This group found a depletion of the green fluorescence, indicative of catecholamine reduction in the caudate nucleus, close to the lateral ventricles, as well as loss of fluorescence in the median eminence and hypothalamic varicosities and again in the neuronal cytoplasm of noradrenergic cells in the locus coeruleus. The dopamine-containing cell bodies of the substantia nigra were not affected. In another study by Uretsky and Iversen (1970), single doses of 6-OHDA at 500 μg/kg produced severe depletion of noradrenaline and dopamine, and the reduction in these two catecholamines was found to be extremely long-lasting, up to 75 days for noradrenaline and 32 days for dopamine, with no knowledge as to whether or not recovery is ever made. Uretsky and Iversen found no changes in the levels of serotonin and γ-aminobutyric acid compared with controls, suggesting that the toxic action of 6-OHDA is selective only for catecholamine neurons in the central nervous system in the same manner that it affects these amines in the periphery. Needless to say, the ability to produce selective lesions of neurochemical origin or chemical sympathectomy by the use of 6-OHDA would be very valuable in the study of neuronal degeneration in the central nervous system. The similarity in chemical composition of 6-OHDA and dopamine cannot be ignored, which leads to the suggestion that in some psychotic reactions, especially schizophrenia, there might be an insufficiency of active dopamine-β-hydroxylase, the enzyme responsible for the conversion of dopamine to noradrenaline, resulting in some release dopamine from nerve terminals that might then be auto-oxidized to toxic 6-OHDA only to be taken up again by the nerve terminals, which it would gradually disintegrate (Fig. 42). Animal experiments by Stein and Wise (1971) showed that after repeated doses of 6-OHDA rats which had been taught to drink milk from a graduated tube lost their interest in resuming this task. Moreover, rats with a permanent indwelling probe inserted into the so-called reward area of the medial forebrain bundle, which the rats will self-stimulate repeatedly for long periods of time, lost their interest in this self-stimulation after the administration of 6-OHDA. Only 200 μg of 6-OHDA could reduce the rate of self-stimulation by 60%–70% of the control values. Further evidence for this theory is that after large doses of chlorpromazine the severe effects seen after 6-OHDA are markedly reduced compared to nontreated controls.

Dual treatment with a MAO inhibitor (pargyline) and 400 μg of 6-OHDA is reported by Stein and Wise to produce severe catatonia in animals. The use of a MAO inhibitor would of course potentiate the dopamine integrity at the synaptic cleft, and if indeed DA is auto-oxidized to 6-OHDA in this region, more of the amine would be available for conversion to its hydroxy analogue, which could then act to destroy the integrity of the nerve terminals. At present, this hypothesis is highly debatable and raises such obvious questions as what is the exact process of auto-oxidation of dopamine in the synaptic cleft. Also, there are the questions

A.

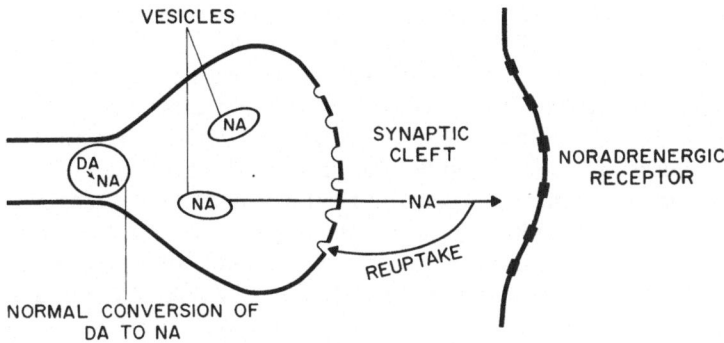

B.

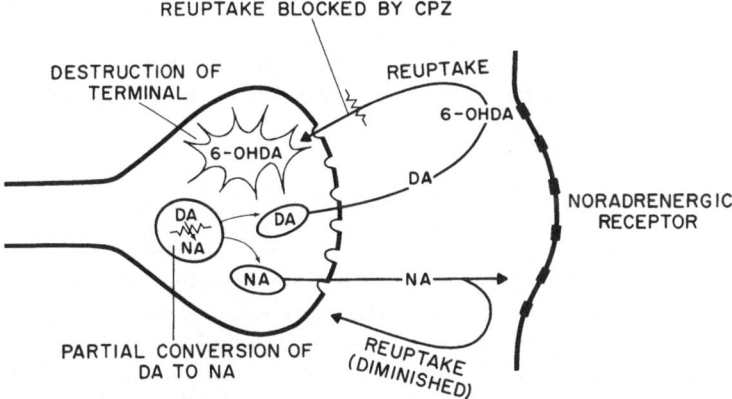

Fig. 42. Nerve terminals in medial forebrain bundle ("reward system") in normal (A) and schizophrenic (B) subjects. The normal release of nor-adrenaline (NA) from vesicles may be metabolically changed through an error in the metabolism of dopamine (DA) to NA, resulting in the release of DA at the synaptic cleft which is abnormally autooxidized to 6-hydroxy-dopamine (6-OHDA), a substance which can degenerate nerve terminals upon its neuronal uptake (B). The process of degeneration may be blocked by prior treatment with chlorpromazine (CPZ). After Stein and Wise (1971).

of whether or not dopamine-β-hydroxylase is deficient in schizophrenics and why, if such a condition is an inborn metabolic error, the symptoms are not seen early in childhood. The problem of regression of schizophrenic symptoms should also be considered, since 33 % of all schizophrenic patients recover regardless of treatment.

More pragmatically, it is perhaps too simplistic to extrapolate from physiological and behavioral reactions in rats in order to form a basis for etiological consideration of schizophrenia in humans. It is also very possible that other 6-OHDA analogues may produce similar effects in these animals, which would rule out the specificity of hydroxydopamine effects on nerve terminals. It should be pointed out that catatonia in rats may be produced by other compounds such as fentanyl-droperidol (Innovar), which causes muscular rigidity after rapid administration. Compounds such as bulbocapnine, reserpine, and chlorpromazine are also capable of producing catatonic-like effects in rats. According to Rogers and Slater (1971), bulbocapnine-induced catatonia does not affect the central concentrations of DA, NA, or 5-HT. In addition, the free and bound fractions of cerebral acetylcholine remain at the same levels following catatonia or postconvulsive seizures in animals receiving reserpine, chlorpromazine, or bulbocapnine. It is possible that 6-OHDA produces the catatonic state reported in rats by Stein and Wise by directly acting on the monoamine metabolism only after subacute or chronic 6-OHDA doses are used. Thus, the dopamine hypothesis remains to be either reinforced or diminished by resolution of these problems, but it is nevertheless exciting and attractive at present to speculate that psychotic disorders may have a neurochemical etiology and that by manipulation of central monoamines these conditions may someday be reversed.

4.2.5. Psychotomimetic Model Psychoses: LSD-25

Drugs which specifically affect behavior and sensory mechanisms in man have been described variously in the literature as hallucinogenic, psychotogenic, psychotomimetic, and, more vulgarly, as "trips." Essentially, these drugs evoke an increase in sensory impulse, enhancing the sense of clarity but diminishing the control over what is being experienced. The most potent of these agents is lysergic acid diethylamide (LSD-25). This compound was discovered by Albert Hoffman, a Swiss chemist, in 1938. Five years later while working with the newly synthesized LSD, Hoffman accidently ingested some of it and a few hours afterward was seized by what he later described as a bizarre mental state resembling schizophrenia. In describing part of that experience, Hoffman wrote, "With my eyes closed, fantastic pictures of extraordinary plasticity and intensive color seemed to surge toward me. After two hours this state gradually subsided and I was able to eat dinner with a good appetite."

The flood of literature that has followed the discovery of LSD could easily fill hundreds of volumes. For our purpose, we shall deal briefly with the actions of LSD on central monoamines as well as with the psychotomimetic action of some phenylethylamines. Briefly, LSD is an ergot alkaloid derived from the plant *Claviceps purpurea,* a fungus which grows on various kinds of grains, particularly on rye. While some of the psychic disturbances induced by ingestion of LSD resemble those of schizophrenia, they are usually reversible within a few hours. LSD has been reported to induce changes in both the catecholamine and serotonin metabolism in the brain. One argument that LSD acts on specific receptor sites in the central nervous system is that of the four stereoisomers synthesized with an LSD structure, only the dextrorotatory isomer (D-LSD) is psychopharmacologically active.

Bromination of the lysergic acid nucleus at the 2-position renders the compound inactive as a psychotomimetic but does not affect its anti-serotonin potency. LSD analogues such as those found in the seeds of *Rivea corymbosa* (ololiuqui) have, for a long time, been used by Mexican Indians for mind-altering effects.

The interrelationship of LSD and serotonin in peripheral structures has already been discussed in Section 4.2.2, but at this point a brief review of the central actions of LSD on serotonin is in order. Freedman and Giarman (1962) reported that rats treated with LSD in doses ranging from 130 to 1500 μg/kg developed a small but statistically significant increase in brain serotonin. This increase in serotonin was found only in the particulate bound fraction after gradient separation. Since these workers could not find any influence of LSD on the 5-HTP decarboxylase enzyme, on monoamine oxidase, or on the brain transport system for 5-HTP, they concluded that the action of LSD is specific at the binding site for serotonin in brain granules. This finding is significant in view of the fact that structurally related analogues of 5-HT or LSD without psychotomimetic action do not affect particle-bound serotonin in the same manner. Microelectrode iontophoresis of LSD in serotonin-containing cells in the raphé nuclei was reported by Aghajanian *et al.* (1970) to produce a slow-down in the spontaneous rate of firing after injection of only 10 μg/kg. Complete cessation in the firing of the cell lasting for about 10 min was produced after a second dose of 10 μg/kg. It should be recalled that these cells in the raphé nuclei are probably the exclusive source of 5-HT-containing nerve cells and that these neurons project via the medial forebrain bundle to the lateral hypothalamus, septal nuclei, globus pallidus, hippocampus, amygdaloid nucleus, and neocortex. Projections from these nuclei have also been found to descend into the spinal cord, as shown by histochemical fluorescence of nerve terminals in this region. If this mechanism of action of LSD on serotonin cells and their firing rate can be accepted as valid, then perhaps LSD could act as a substitute neurotransmitter for serotonin, thus altering and amplifying sensations while distorting the transfer signals from the axon tip across the synapse. The result might then be interference with thalamic cortical pathways, pallidocortical pathways, and hypothalamic pathways, causing motor, sensory, and psychic disturbances.

However, this may be a simplification of even more complex mechanisms, since it has also been found that LSD can affect the noradrenaline turnover in the brain, resulting in a decrease in the central levels of this amine. Rats treated with a MAO inhibitor followed by 0.5 mg of LSD were found by de la Torre (1968a) to have increased nerve terminal fluorescence in the lateral and medial regions of the septum but nowhere else. More interesting yet, catecholamine-containing *cell bodies* in this region were visualized using the histochemical fluorescence method (Fig. 43). Cell bodies in this region containing catecholamines have not previously been described even after strong doses of a MAO inhibitor. These effects could not be reproduced with BOL-148, the psychoinactive isomer of LSD. A possible explanation for this phenomenon could be increased turnover of noradrenaline by LSD which is visible only because MAO has been inhibited before it has had the opportunity to deaminate the newly released noradrenaline. This would also explain why this effect is hardly visible with single injections of LSD without prior MAO inhibition. It will remain to be seen whether other psychotomimetic substances produce the same effects.

The actions of LSD can be enhanced in normal human volunteers when reserpine is previously administered. By contrast, use of a MAO inhibitor with LSD diminishes its psychotomimetic action (Resnick *et al.*, 1967). It will be recalled that MAO inhibitors are known to increase and reserpine to deplete biogenic amines in the brain, particularly the catecholamines. Treatment of normal subjects with L-tryptophan, a serotonin precursor, was reported by Resnick *et al.* (1967) to significantly attenuate the effects of LSD in all of them. In view of the blood–brain barrier to 5-HTP and L-dopa, it would be of interest to see whether pretreatment with dopa decarboxylase inhibitor followed by the monoamine precursor can attenuate or increase the effects of LSD in humans. There is probably a blood–brain barrier to LSD, although not as great as that to mescaline. For one thing, LSD is about 70 times more lipid soluble than mescaline. Lipid solubility, it will be recalled, is an important factor that determines the relative penetration of a substance into the brain tissue.

One look at the structural formula for mescaline (Fig. 44) is enough to send one into protracted speculation on the possibility that some psychoses can be produced by abnormal methylation of the catecholamines. Indeed, mescaline resembles dopamine except for the 3-methoxy groups at the 3-, 4-, and 5-positions. This speculation is further reinforced by the observations of Pfeiffer *et al.* (1967) that (1) methionine and other methyl donors increase the degree of psychosis; (2) primary amines do not readily pass the blood–brain barrier unless they are alkylated to make them more lipid soluble and increase their passage into the brain parenchyma; (3) methylation of these amines increases their central nervous system stimulant effect; (4) this stimulant effect on the central nervous system is a charac-

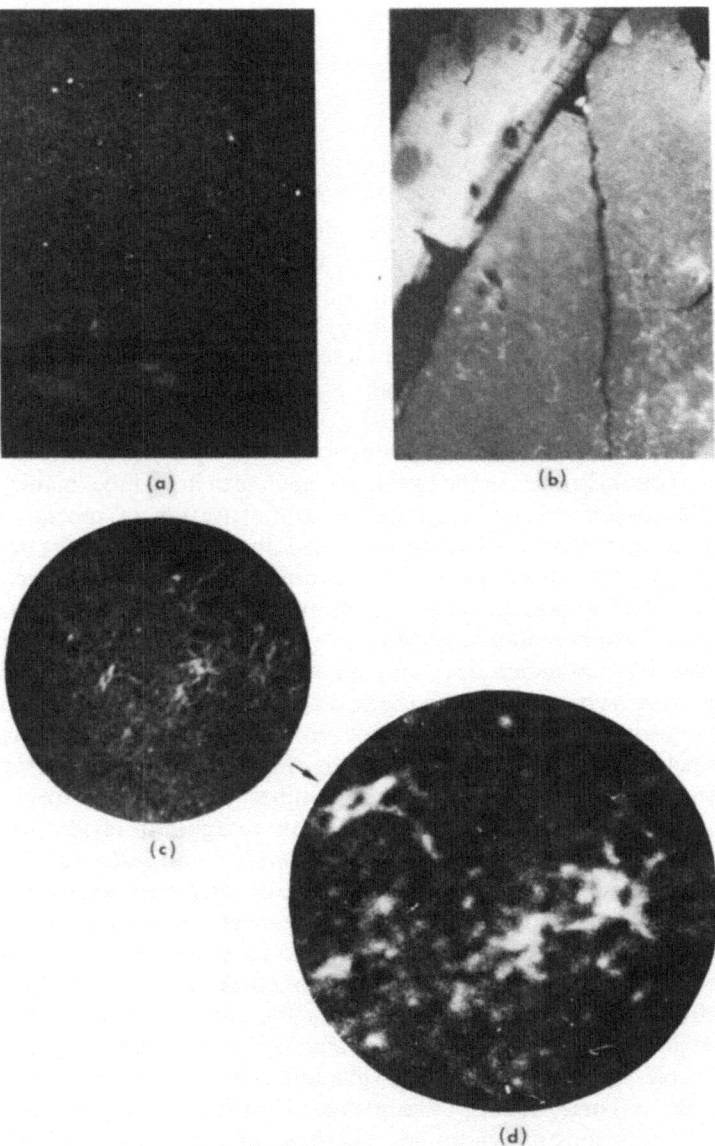

Fig. 43. Effects of nialamide plus BOL-148 on the septal region of rat brain. Normal terminal fluorescence (a) resembling nialamide treatment only. After nialamide plus LSD-25, increase of terminal fluorescence intensity and appearance of catecholamine-containing cell bodies (b, c, d) are noted in this region. Caudate nucleus (b) is seen at top left separated from lateral septal region by lateral ventricle. Blow-up magnification (d) shows cells and their processes with typical nonfluorescent nucleus in mediolateral septum. 80×. (Reduced for reproduction 40%.)

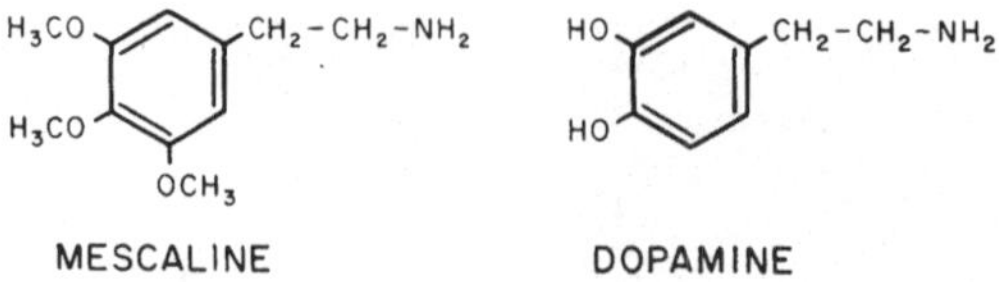

MESCALINE   DOPAMINE

Fig. 44

teristic of all psychotomimetic drugs; and (5) chronic schizophrenics appear to have a continuous arousal of brain waves shown by EEG as though a potent central nervous system stimulant were in action. Psychotomimetic action is not dependent on a methylated phenylethylamine structure. It can also occur in a methylated indoleamine such as dimethyltryptamine and bufotenine. Both these substances are powerful psychotomimetics in man.

We have reviewed how psychotropic drugs can mobilize the concentrations of monoamines in the brain. We have seen how direct manipulation of the blood–brain barrier (by direct administration of precursors) can increase either the catecholamines or serotonin. The psychotropic drugs can have unique or multiple actions on the monoamines. They can increase or decrease catecholamines without affecting serotonin; they can increase or decrease serotonin without affecting the catecholamines; they can increase or decrease *all* monoamines; and, in some cases, they can increase or decrease *specific* monoamines in the brain. However, none of these combinations has proven psychotogenic in action. It is this consequence that leads one to believe that psychotogenic activity in the central nervous system is, if indeed it can be related to the monoamines, probably site specific; that is, with a substance like LSD we can see an antagonism of serotonin concomitant with increased indoleamine levels in the brain and at the same time a small decrease in the concentration of noradrenaline with little or no effect on the dopamine concentration. The problem is far from resolved. Autoradiographic studies not yet confirmed have shown that radioactively labeled LSD probably deposits in the hypothalamus, an area rich in noradrenaline and serotonin. It is clear that the action of LSD would make more sense if it specifically mobilized amine concentration at specific sites, for example, in the raphé nuclei (Aghajanian *et al.*, 1970) or in the septal region (de la Torre, 1968a). The problem is further compounded by the fact that it is impossible in animal experimentation to know what, if any, psychic changes are taking place after the administration of psychotomimetic compounds. The answer may have to wait until a more refined technique is available for the determination of monoamine concentrations in the human brain at autopsy of psychotic subjects. Until that time, our model chemical psychoses will remain an Elysian mystery or a psychodynamic challenge, depending on how one looks at the matter.

4.3. NEUROPATHOLOGICAL STATES

4.3.1. Neuroendocrine States

There is probably no biological activity which is not influenced by cyclic variations or rhythms. That which we call diurnal or circadian rhythm constitutes one of the fundamental aspects of life. The simplest example of this is cell division. The source of the regulation of these biological clocks can be sought within the subcellular or cellular structure. Our purpose in discussing circadian rhythms in neuropathological states is to review how monoamines have been implicated in various neuroendocrine disorders.

It is still not known what governs or modulates the delicate balance of biological rhythms or how external or internal forces can disrupt this balance. The role of the central nervous system in cyclic endocrine functions is closely linked to circadian rhythms and pituitary adrenal functions. It is probable that the hypothalamus is the most involved cerebral structure influencing the secretory activities of the anterior lobe of the pituitary. This does not imply that the latter is completely under hypothalamic control, because in accordance with changes in the concentration of various hormones in the blood one must assume that a great number of anterior lobe functions are autonomous. The hypothalamus is a small basal structure which in man weighs approximately 4 g; it is situated at the floor of the brain and alongside the walls of the third ventricle, flanked laterally on each side by the internal capsule and bordered anteriorly by the optic chiasm and posteriorly by the interpeduncular fossa, while the thalamus serves as the roof. The boundaries of the hypothalamus are of considerable importance in the diagnosis of hypothalamic lesions, since they frequently involve vegetative, endocrine, thalamogenic, and pyramidocerebellar signs when adjacent structures of the hypothalamus are involved. The pituitary gland is connected to the hypothalamus by a stalk. Within the stalk is found the supraopticohypophyseal tract, which serves as the neurosecretory route to the posterior lobe of the pituitary gland from the hypothalamus, as well as hypothalamohypophyseal portal system, which links the two structures with a network of blood vessels that supply the parts.

The pituitary is innervated mainly from the supraoptic and periventricular nuclei and the tuberoinfundibular nucleus found in the hypothalamus. These nerves enter the median eminence through the main bundles of the hypophyseal tract and are thought to contain dopaminergic fibers. The cell bodies which are visible after histochemical fluorescence appear to originate from the tuberoinfunidbular region. Osmosensitive neurons in the supraoptic nucleus have been identified by recording the increase in rate of discharge following intra-arterial injections of sodium chloride solutions.

After hypothalamic stimulation, all osmosensitive cells increase their firing, but no relationship is found between the firing rate and the blood pressure changes following or during stimulation of the reticular formation, the thalamus, or the hypothalamus. The supraoptic nucleus receives an exceptionally rich blood supply, and changes in the activity of its neurons seem to be caused primarily by variations in the osmotic pressure of the blood. When hypothalamic vessels become hyperosmotic—for example, after dehydration—the supraoptic nucleus cells respond by accelerating the formation and/or release of antidiuretic hormone (ADH). When blood with a high titer of ADH reaches the kidney, the hormone probably acts by controlling the reabsorption of water in the distal convoluted tubules through blood osmotic pressure. There is good experimental evidence to suggest that ADH is a neural secretion produced by the supraoptic nucleus neurons. The neurosecretion is thought to migrate through the nerve fibers of the supraopticohypophyseal tract to be released into the neural (posterior) lobe capillaries in the hypophysis. The relationship of diabetes insipidus to neurons of the supraoptic nucleus can be seen by the almost complete loss of cell bodies in the supraoptic region and their axons in the infundibulum and posterior hypophyseal lobe. Besides ADH, the supraoptic nucleus is also responsible for the elaboration of vasopressin, a polypeptide hormone that increases blood pressure and decreases urine flow through neural lobe secretion. The periventricular nucleus is concerned with developing oxytocin, a hormone involved in the contraction of uterine muscle and milk ejection in the female. Reports indicate that suckling predominantly releases oxytocin and dehydration releases mostly vasopressin. Bladder contractions and inhibition of peristalsis can be evoked by stimulation of the lateral hypothalamus. The former effect is believed to be associated with the group of fibers of the medial forebrain bundle that ascends from the mammillary bodies to the periaqueductal gray matter. The latter effect is thought to be a function of the tegmental component of the medial forebrain bundle.

Scott and Pfaffman (1967) have shown that electrical stimulation of the rat olfactory bulb can elicit discharges in the medial forebrain bundle region of the lateral hypothalamus. This olfactory input appears to be of importance in feeding and sexual behavior, at least in rats. Male rats prefer the odor of estrous females to nonestrous, and mating ability in the male is reduced following olfactory bulb ablation. Odors also have a marked effect on pregnancy, endocrine functions, and aversion to foods. Preoptic electrochemical stimulation is reported by Everett and Quinn (1966) to produce pseudopregnancy in rats when accompanied by stimulation of the tuberal region. When only tuberal stimulation was applied, corpus luteum formation and the start of pseudopregnancy were delayed for 24 hr. An arousal mechanism involving normal hibernating animals may be associated with the preoptic anterior hypothalamus. Satinoff (1967) observed normal hibernation in ground squirrels after preoptic region damage, but death instead of arousal occurred at the end of the hibernation period. The lesion may equally have caused damage to the thermoregulatory center in the hypothalamus. The ventromedial nucleus has been implicated by numerous studies in the mechanism regulating feeding behavior, and a "satiety" center has been proposed for this nucleus by many authors.

An unsettled question concerning the function of the hypothalamus and neuroendocrine control is whether each nuclear configuration is responsible for a particular integrated activity or whether there exist two hypothalamic zones as related to autonomic action, suggested by Hess (1956). Hess divided the hypothalamus into two anatomofunctional zones, the ergotropic zone, responsible for sympathetic neurovegetative signs, and the trophotropic zone, concerned with inhibition of parasympathetic neurovegetative signs. As stated before, Brodie *et al.* (1966*a*) suggested that this hypothetical zonation is related to serotonin in the case of the trophotropic zone, involving slowing the heart rate, dilating the blood vessels in stomach and intestines, and mediating those events requiring processes of repair. Catecholamines, on the other hand, would be related to the ergotropic zone in such sympathetic activities as accelerating the heart beat, increasing blood pressure, and mediating such responses as aggression and rage. Gonadotropin secretion has been suggested by Fuxe and Hökfelt (1969) to be regulated by at least two monoamine systems, for example, that represented by the noradrenaline afferents to the hypothalamus and the other by the tuberoinfundibular dopamine neuron system. This would imply a dopaminergic pathway at the level of the median eminence and infundibular stem. Indeed, histochemical fluorescence has confirmed findings by Fuxe and Hökfelt (1966) of fluorescent terminals surrounding the primary plexus of the hypophyseal portal system. Cervical sympathectomy indicates that these terminals may be fed by central neurons originating from catecholamine-containing nerve cells in the region of the tuberoinfundibular tract. The terminals may contain chiefly dopamine since fluorescence is obtained in this region after 24 hr previous to administration of α-methylmetatyrosine, a compound that is known to retard noradrenaline fluorescence for several days. However, there is a possibility that noradrenergic fibers may also be involved in this system.

Recent evidence has focused attention on the role of dopamine localized in the tuberoinfundibular region of animals in various endocrine changes, especially in pregnancy. Work by Fuxe and Hökfelt (1966) suggests that dopamine-containing neurons in the infundibulum show some changes in the dopamine turnover correlated with gonad function and with sex hormones. Thus, early pregnancy was associated with an increase in number and intensity of dopamine-containing cell bodies in the arcuate nucleus.

This increase was maintained throughout the gestation period. The same phenomenon was observed during pseudopregnancy and during the 2–3 weeks of the lactation period following birth. In addition, the dopamine-containing nerve terminals in this region appear to be related to the estrus cycle in the sense that the dopamine turnover rate is diminished during the ovulation period. No modification of the cycle was noted after gonadectomy or following androgen sterilization of these rats. More recently, Fuxe and Hökfelt (1970) have shown that prolactin appears to activate the tubero-infundibular dopamine neurons, an effect which is not seen with follicle-stimulating hormone or luteinizing hormone. The conclusion by these authors is that prolactin may be activating the tuberoinfundibular neurons during pregnancy, pseudopregnancy, and lactation. The speculative implications of these findings are extremely important to our understanding of ovulatory mechanisms and systemic hormonal balance. The possibile selective manipulation of dopamine in the infundibulum in order to control sustained ovulation for long periods, especially in these times of concern about demographic population explosion, is a research area ripe for immediate exploration.

Kamberi and Kobayashi (1970) have reported that the activity of monoamine oxidase changes cyclically in the hypothalamus, amygdala, adrenal, and ovaries. These workers found that during the estrus cycle, much lower levels of activity of MAO were evident in other regions of the brain, that is, in the frontal and lateral cerebral cortex, as well as in the pituitary and in the uterus, and none of these showed any cyclical changes. There are also data the the luteinizing hormone releasing factor (LHRF) may be controlled by dopamine in rats. Schneider and McCann (1970) presented data indicating the existence of dopaminergic transmission for the LHRF secretory neurons. The sexual cycle has also been connected with noradrenaline, and reports by Kurachi and Hirota (1969) show that significant changes in diencephalic noradrenaline content occur in the sexual cycle, after oophorectomy, and after administration of estrogen following oophorectomy. However, when Segal and Whalen (1970) compared the effects of estrogen and progesterone with the effects of estrogen plus p-CPA, the depleter of brain serotonin, on the induction of sexual receptivity, they found that progesterone facilitated the effects of estrogen on the induction of receptivity. They concluded, therefore, that no direct relationship existed between brain serotonin and receptivity. Systemic administration of dopamine to rats was found by Leonardelli (1969) to cause an interruption of the luteinizing hormone. This result can be related to the fact that monoaminergic neurons of the ventral tuber are hyperactive in guinea pigs during the preovulatory period, in the course of which the gonadotropic functions are simulated.

The function of serotonin levels in the pineal body has also been investigated in relation to circadian rhythms. Control of this rhythm may be disrupted by interrupting the communication of postganglionic fibers originating from the superior cervical ganglion. Even though the sympathetic innervation in the rat is not fully developed 5 or 6 days after birth, there still exists a 24 hr serotonin rhythm in the immature rat pineal gland. Serotonin rhythm in the adult pineal gland can be abolished by constant light, but it continues in total darkness (Schneider *et al.*, 1965). Histochemical fluorescence and electron microscopy of the pineal show that 2–3 weeks are required for the gland to become completely innervated. Enzymes such as hydroxyindole-*O*-methyltransferase (HIOMT) appear to be involved in the *O*-methylation of *N*-acetylserotonin. Also, 2,4-melatonin appears to be controlled by environmental light. Studies involving this phenomenon are continuing, and thus far it would appear that central lesions may affect enzymatic levels in tissues not directly innervated by the nerve cells destroyed, such as in the pineal. This phenomenon has been termed "trans-synaptic neurochemical effects after central lesions." It appears from the above data that hypothesis of adrenergic mediation in the regulation of adrenocorticotropic function of the hypophysis would have some merit. The assumption that dopamine or noradrenaline may mediate some of these neuroendocrine reactions appears very likely as more evidence is presented.

We have already examined in Section 3.5 how noradrenaline and serotonin may mediate various phases of sleep activity, as well as hibernation. These normal diurnal cycles in both animal and man could conceivably be easily disrupted by physical or chemical means. Other conditions, such as temperature, drinking, and feeding mechanisms, may also be regulated by monoamines. Basing their concept on experiments using intraventricular and local injections of monoamines, Feldberg and Myers (1963, 1964) have proposed that thermoregulation is mediated through the release of endogenous monoamines. Their conclusion is that a function of serotonin is to control heat gain in the organism, while catecholamines are involved in heat loss. This concept again is allied to Hess' ergotropic and trophotropic amine behavior mechanisms. It is well known that after destruction of the hypothalamus most mammals lose their ability to maintain a uniform body temperature. In man, many cases have been observed where tumors or other hypothalamic lesions have changed the homothermic state into a poikilothermic condition. Many neurons susceptible to increases in blood temperature are situated in the anterior region of the hypothalamus, possibly extending into the preoptic area. These neurons participate in the mechanism whereby excess heat is released. Their descending pathways join in the central gray structures in the midbrain and set up connections with the cardiovascular and respiratory mechanisms located in the brain stem and spinal cord. The neurons are affected by blood temperature changes and are scattered in the hypothalamus, mostly below the supraoptic region, where they have similar connections with the brain stem. Thermosensitive elements in the hypothalamus are normally cooled by regional blood vessels, and it is not too difficult to imagine that changes in arterial blood temperature can initiate thermosensitive reactions in the hypothalamus if such regional blood flow is altered. Thermoregulatory mechanisms, however, appear simpler than they really are. Rawson and Hammel (1963) suggest that hypothalamic temperature could be a function dependent on four parameters: (1) temperature of blood flowing through it, (2) rate of blood flow, (3) rate of heat production, and (4) heat dissipated to surrounding regions.

A fall in temperature in anterior and posterior hypothalamus has been described by McCook *et al.* (1962) following small intraventricular injections of sodium pentobarbital, which they attribute to increased hypothalamic blood flow. Goodman and Gilman (1955) suggest that anesthetics suppress temperature regulation so that greatly magnified deviations in body temperature are necessary to affect thermoregulatory responses. Sleeplike conditions have been shown by Feldberg (1958) to result after local intraventricular administration of anesthetic-like agents into conscious cats. Evidence thus would support the suggestion that serotonin injected into the lateral ventricles of cats tends to produce an elevation in temperature of the animals, while noradrenaline injections cause the temperature to fall. Similar results were observed in the dog and in the monkey (Feldberg *et al.*, 1966). Both noradrenaline and serotonin are known to exert strong effects on cerebral blood vessels. Monoamine influence on hypothalamic blood flow has not yet been measured, but it is probable that when it is, a difference in effects will be found. Two propositions are suggested implicating the hypothalamus as a thermoregulatory center: (1) that normal body temperature may be determined by relative quantities of monoamines released in the hypothalamus and (2) that fever caused by various pyrogens may result through excess serotonin released in the hypothalamus.

Two points will need to be clearly demonstrated if the above suggestions are to be accepted: (1) the action of monoamines on the thermoregulatory mechanism and (2) the regulation of body temperature by monoamines. Drinking and feeding mechanisms constitute in the animal a primary segment of behavior. Lesion or stimulation of certain hypothalamic regions can elicit either excessive drinking (polydipsia) or feeding (hyperphagia) in most animals. Thus, local feeding centers have been described for the lateral hypothalamic region, while a satiety center is responsible for reduced food intake, but not complete aphagia. Grossman (1962), using double-walled cannulas, introduced minute amounts of noradrenaline into an area between the mammillothalamic tract and the fornix, corresponding to the lateral and dorsal aspects of the ventromedial nucleus. He noted that noradrenaline induced in food-satiated rats an increase in eating and a decrease in drinking behavior. If noradrenaline was replaced by acetylcholine, the reverse action was obtained. This and other findings led him to suggest that the ventromedial area contains two cell populations able to regulate food and water intake that are preferentially sensitive to cholinergic and adrenergic stimulation. In addition, he found that various cholinergic and adrenergic blocking agents could affect the neurochemical property of these hypothalamic neurons by reducing normal hunger and thirst, respectively. Consequently, the hypothesis has been offered that feeding is mediated by an adrenergic mechanism and drinking by a cholinergic mechanism. Neuroanatomical studies by Arees and Mayer (1967) have shown a direct fiber connection between the ventromedial hypothalamic nucleus (so-called satiety center) and the lateral hypothalamic area (so-called feeding center). These two workers found that axons from the ventromedial nucleus extend to the lateral hypothalamic nucleus and could act as an inhibitor of the latter structure. Coury (1967) investigated the interrelations of combined adrenergic and cholinergic stimulation on the Papez circuit (the neural system proposed for emotional activity) and found evidence to corroborate previous studies indicating that eating and drinking behavior are elicited by adrenergic and cholinergic site stimulation, respectively. On the basis of his findings, he proposed that separate synaptic systems are deployed throughout the brain and cerebral cortex to control hunger and thirst. Noradrenaline has been reported by Booth (1967) to produce eating in rats when injected in microquantities into regions other than those parts of the lateral hypothalamic nucleus normally associated with feeding. It is still not certain whether the hypothalamus is important in food and water intake merely as a point of convergence for numerous nervous pathways or whether it is directly concerned with the integrated functions of these two mechanisms. The picture that emerges from the studies reviewed supports the theory that the hypothalamus is integrative; however, final evidence is still lacking.

What exactly the effects of tumorous destruction of hypothalamic structure are on behavior and diurnal cycles is an issue that has not received much research attention. In both animals and man, electrical stimulation of the hypothalamus can produce behavioral reactions. Destruction of specific hypothalamic regions such as the ventromedial nucleus can produce in the animal rage and aggressive behavior. It is possible that certain tumors in humans may damage the hypothalamic areas concerned with food regulation, resulting in loss of appetite and subsequent severe emaciation, as in Simmond's disease. Conversely, adiposogenital dystrophy, as manifested in Fröhlich's syndrome, could result from destruction of the inhibitory food regulation center, leading to greater than normal obesity. It is well known that patients suffering from head injuries, cerebroarterial sclerosis, or disseminated sclerosis frequently display emotional instability and increased behavioral reactivity, characterized by angry outbursts and compulsive crying or laughing fits. These symptoms may result from impaired control of the posterior hypothalamus by higher centers, although it should be pointed out that the complex mechanism of emotions may be assumed to include the hypothalamus and also thalamocortico- and thalamohypothalamic connections. Under certain conditions, therefore, aggressive psychopathological behavior may have as its source specific nerve cell necrosis and monoamine alterations due to cerebral trauma involving in some way hypothalamic receptor sites. The conditions affecting traumatic brain injury will be discussed in Section 4.3.3.

4.3.2. Experimental Seizures

The idea that serotonin may be related to seizures was first advanced by Bonnycastle *et al.* (1957), who showed that many anticonvulsants such as diphenylhydantoin, phenobarbital, and nitrazepam increase brain serotonin concentrations when given to animals at optimum doses. This serotonin increase was not seen in organs other than the brain. It is well known that in subjects with a dietary deficiency of pyridoxin and in patients treated with isoniazid, neurological disturbances such as polyneuritis, mental confusion, and convulsive attacks may develop in addition to disturbances of the digestive apparatus, the skin, the mucosa, and the blood. In reference to the convulsive attacks, various pathogenic theories have been advanced. The principal theory is based on the effect of pyridoxin on the metabolism of the brain and especially on the role of pyridoxin in the production of serotonin. Some beneficial effects using vitamin B_6 have been reported in epileptic children, who showed a consistent high excretion rate of xanthurenic acid, a tryptophan metabolite. Epilepsy produced in monkeys and cats by means of aluminum hydroxide has been reported by Wada (1961) to have some relationship to brain serotonin levels. Monoamine oxidase inhibitors, which elevate the brain levels of serotonin and noradrenaline, in contrast to reserpine, which lowers these levels, have been shown to prevent the tonic extension of hindlimbs of rats produced after an electric shock. These MAO inhibitors included iproniazid and phenylpromazine, while isoniazid, which does not inhibit MAO, showed no anticonvulsant effect. Thus, the anticonvulsant properties of MAO inhibitors offer an opportunity to study some of the actions of MAO and its substrates in the brain. Since reserpine reduces brain amines and enhances convulsions, while MAO inhibitors increase the amines and protect against seizures, it seems possible that certain physiologically active amines may be important in determining the sensitivity of brain to experimentally induced seizures. This suggests that certain types of epilepsy may involve a localized dysfunction in the formation and release for metabolism of monoamines. It seems likely then that injection with the serotonin precursor 5-hydroxytryptophan would offer some degree of protection to animals made convulsive by the use of various drugs such as pentylenetetrazol (leptazol). This, indeed, has been tried by Chen et al. (1968) and by Minor (1968), but the end result did not meet the theoretical expectations. In mice, for example, 100 mg/kg of 5-HTP had no effect on the extensor seizure threshold following electric shock, although 300 mg/kg of p-CPA, the serotonin depleter, did lower the resistance of the mice to such seizures. It is worthwhile to recall that there exists a blood–brain barrier to 5-HTP also, mainly peripheral dopa decarboxylase. De la Torre and Mullan (1970) have reported that treatment with the peripheral dopa decarboxylase inhibitor Ro 4-4602 followed by 5-HTP markedly protects rats from leptazol-induced seizures (Table 16). This degree of protection was seen to a lesser extent when phenobarbital was used as a control drug. It was also confirmed that the brain serotonin depleter p-CPA increases the severity of the drug-provoked convulsions (Table 17). Disulfiram, which acts to inhibit the formation of noradrenaline in brain and hence to decrease the level of this amine, had no apparent effect on the seizure threshold. These findings suggest that increasing 5-HT through peripheral decarboxylase inhibition and 5-HTP administration might result in some type of inhibition of neuronal excitability induced by leptazol. This response is not evident when 5-HTP is given alone, because most of the amino acid is quickly

Table 16. Relative Seizure Susceptibility to Leptazol After Various Drugs[a]

Drug	Dose (mg/kg)	Phase I		Phase II		Phase III		Phase IV
		A	B	C	D	E	F	G
Disulfiram[b]	3 × 100	—	—	10 (1)	10 (1)	80 (8)	—	—
α-Methyltyrosine[b]	3 × 100	—	—	10 (1)	20 (2)	70 (7)	—	—
Ro 4-4602[b]	50	—	5 (1)	5 (1)	5 (1)	70 (14)	—	15 (3)
5-HTP[b]	100	—	5 (1)	30 (6)	5 (1)	60 (12)	—	—
Phenobarbital[c]	6	10 (2)	5 (1)	55 (11)	30 (6)	—	—	—
Ro 4-4602 + 5-HTP[d]	50 + 100	35 (7)	65 (13)	—	—	—	—	—
Leptazol	40	—	—	10 (8)	5 (4)	80 (64)	2.5 (2)	2.5 (2)

[a] Phase I indicates no or very mild seizure, phase II moderate seizure, phase III severe convulsions, phase IV death. Numbers indicate percent of rats; total number of animals is shown in parentheses. There comparatively good protection in phenobarbitone-treated rats (75% under phase II), while excellent protection is seen after Ro 4-4602 plus 5-HTP (100% phase I).

[b] $P > 0.05$ (ns).

[c] $P < 0.01$.

[d] $P < 0.001$.

**Table 17.   Seizure Susceptibility in p-Chlorophenylalanine (p-CPA)
Treated Rats[a]**

Drug	Dose (mg/kg)	Phase I		Phase II		Phase III		Phase IV
		A	B	C	D	E	F	G
p-CPA[b]	316	—	—	—	—	—	100 (20)	—
p-CPA[c]	3 × 100	20 (4)	—	—	15 (3)	25 (5)	40 (8)	—
Leptazol	35	45 (9)	—	25 (5)	—	35 (6)	—	—

[a] Leptazol was reduced to 35 mg/kg for better evaluation of the p-CPA effect. Rats treated for 3 days with 100 mg/kg daily of p-CPA show a decrease in seizure threshold, with 65% in phase III, severe convulsions. Susceptibility to leptazol is markedly enhanced after a single strong dose of p-CPA as seen by 100% phase III, severe convulsions, versus 70% leptazol controls in phases I and II. From de la Torre and Mullan (1970).
[b] $P < 0.001$.
[c] $P < 0.01$.

transformed to serotonin at the capillary level, remaining outside the blood–brain barrier.

The electrolyte balance in seizures also appears to be important. It is known that in convulsive seizures potassium flows out of nerve cells and excessive sodium flows in. Schneider and Thomalske (1963) found that the retention of sodium can lead to relative depolarization which generally exacerbates or induces spontaneous electrochemical discharges in centrencephalic seizures. They also found that adrenocorticotropic hormone (ACTH) improves the electrolytes on both sides of the cell membrane, consequently correcting the electrolyte metabolism, which they believe is a factor in centrencephalic seizures. These findings are of interest in the light of a recent investigation by Curzon and Green (1968) linking ACTH to selective release of serotonin from brain nerve cells. Drug-induced seizures produced by leptazol can also cause acute swelling of the astrocytes, which by compression may impair cerebral blood flow and could conceivably result in alteration of the blood–brain barrier permeability to various electrolytes. This effect has been reported by De Robertis *et al.* (1969) and confirmed using electron microscopy by de la Torre (unpublished report). Areas that remain to be investigated are those concerning the action of central serotonin, involving the regulation of intracellular sodium and potassium. It is not yet known whether reduced levels of brain serotonin produced chemically or mechanically can increase the depolarization of neurons and therefore affect the cell electrolyte metabolism. The principal question is whether or not serotonin, adduced to be an inhibitory substance in the central nervous system, can reduce neuronal excitability.

The association of electrolyte imbalance and neuronal physiological function cannot be overstressed. Eccles (1964) showed that potassium is released into the extracellular space during neuronal activity. If the potassium is increased extracellularly, neuronal depolarization leading to an increase in excitability at the presynaptic terminals occurs. In the normal brain, the potassium is transported away from its site of release so that marked neuronal excitability is prevented. Thus, epileptogenic neurons, resulting in prolonged depolarization and hyperpolarization with spontaneous and repetitive firing, may reflect an impairment of the cell membrane permeability or a malfunction in the potassium transport system close to the membrane. Precisely how this ionic transport system ties in with the monoamines is as yet unclear. It is also of interest to note that after acute leptazol-induced seizures, electron microscopy reveals astrocytic swelling around the brain capillaries. Depolarization of neurons induced by the leptazol may result in increased extrusion of potassium, which is taken up by the astrocytes. In turn, the swelling of the astrocytes may reduce blood flow, causing ischemia or, if severe enough, death. Besides, it has been shown by Tower (1960) that slices of brain scar tissue from epileptic patients do not take up potassium and do not release sodium as normal brains do. Seizures in cats may also be produced after intraventricular injections of potassium or sodium. The increased intracellular sodium after adrenalectomy increases cerebral excitability (Timiras et al., 1954), while anticonvulsant drugs such as diphenylhydantoin and acetazolamide appear to decrease intracellular sodium. Bogdanski and Brodie (1969) have advanced the notion that inorganic electrolytes are important in the regulation of the storage and accumulation of noradrenaline in sympathetic nerve endings. Since manipulation of brain serotonin but not noradrenaline (de la Torre and Mullan, 1970) has been seen to modify drug-induced seizures, it is tempting to speculate on a similar action by potassium in the active transport and storage of serotonin, while sodium would be concerned with the binding of serotonin and with preventing its spontaneous release from its storage site. Preliminary evidence (de la Torre and Lim, unpublished report) indicates that after severe head injury in cats produced by extradural compression with inflated balloons, the sodium and potassium ATPase activity is increased in the compressed hemisphere compared to that in the contralateral hemisphere. In rats, with leptazol-induced seizures, the typical astrocytic swelling seen around brain capillaries appears diminished after endogenous elevation of serotonin through peripheral decarboxylase inhibition and 5-HTP administration (de la Torre, unpublished results).

4.3.3. Brain Trauma

Osterholm et al. (1969) reported that minute quantities of serotonin injected intracerebrally in cats resulted in severe neurological alterations.

This study was followed by some additional reports by the same authors, who theorized that free serotonin displaced from cells by physical means could act upon the reticular formation to produce coma. The same free serotonin in medullary cardiorespiratory structures would cause death. Besides that, they pointed out that serotonin in extracerebral tissues is extremely effective in producing edema. For example, 1 μg of serotonin injected into subcutaneous tissue in the rat can retain up to 2 ml of water. After injecting 50–200 μg of serotonin in the white matter of cats, close to the ventricular ependyma, Osterholm's group found that the resulting neurological deficits permitted classification of the animals into three groups. Group 1, which they considered to have mild neurological deficits, as characterized by somnolence and decreased reactivity to painful stimuli. Walking was difficult, and a mild contralateral hemiparesis was found in a few animals. Group 2 they described as having severe neurological deficits. These animals suffered from hemispheric or brain stem dysfunction following serotonin injections. The animals had generalized convulsions, were profoundly hypothermic (76 F), and had respiratory rates as slow as 5 per minute. About 34% of these animals were found to be decerebrate with extreme degress of opisthotonus. The majority of the animals in this group were severely comatose and unresponsive to any stimulus and had fully dilated pupils and a contralateral hemiparesis. The third group of animals died within 12–24 hr after the intracerebral serotonin injection. If crystalline serotonin on gel foam was placed near the ventricular wall, about 50% of the animals suffered severe neurological alterations, while 60% died after 24 μg of crystalline serotonin was injected into the brain. These studies suggested that free serotonin in contact with cerebral tissue induced three basic manifestations: (1) cellular excitation characterized by cerebral seizures; (2) cerebral edema, and (3) acute depression of the neural function resulting in hemiparesis, decerebrate rigidity, coma, and death. In view of the fact that serotonin is capable of producing striking effects on nerve activity after microelectrophoretic application and can depress either spontaneous or induced activity, these reports are, indeed, of interest. We have reviewed thus far how serotonin may affect slow sleep activity, temperature regulation, sexual and other endocrine reactions, convulsive seizures, and various psychogenic states. If the reports on serotonin-induced cerebral edema are confirmed, it could greatly add to our knowledge of traumatic head injury, which affects about 50,000 people annually, with the likelihood that an effective therapeutic agent such as a serotonin antagonist could counteract or reverse the neurological consequences of this amine. The idea that serotonin can be released from nerve cells after traumatic injury and in its free form stimulate central receptors or other neurons, which in turn produce various neurological deficits or death, is theoretically acceptable.

However, a recent study by de la Torre (unpublished results) on eight cats and four monkeys failed to confirm Osterholm's results. Subcortical injections of serotonin in cats failed to produce any neurological deficits, and the only gross behavioral alterations seen were the animals' aversion to water (adipsia) and food (dysphagia) coupled with increased somnolence for about 3 days. No changes in the EEG, intracranial pressure, blood pressure, temperature, or respiration were observed in any of the animals, even after bilateral intracerebral injections of serotonin in doses exceeding 300 times those reported by Osterholm. Monkeys treated in this manner did not show any depression of eating or drinking and were completely normal postsurgically.

Serotonin has also been implicated in migraine attacks in humans. It is well known that certain foods give rise to episodes of headaches and hypertension in patients receiving MAO inhibitors. Such foods are always rich in amino acids such as tyramine, histamine, and butaphenylethylamine, which exert a hypertensive effect on the circulatory system and are potentiated by MAO inhibitors. The same foods may provoke attacks in migranous subjects. Serotonin has been mentioned as one of the factors causing migraine, since it exerts a powerful vasoconstrictor effect on the superficial temporal artery. It has been suggested that migraine may be due to a fall in blood serotonin levels such as produced by an injection of reserpine. Recent reports have shown that attacks of migraine may be arrested by injecting serotonin or serotonin antagonist (viz., methysergide) or even MAO inhibitors, which raise the levels of serotonin in the brain. However, the principal source of pain in migraine is the dilatation of the cranial arteries, which may be counteracted by the vasoconstricting properties of serotonin. While there do not seem to be any effects on the parasympathetic and sympathetic system during migraine attacks, there is a change in the neuroreflex control of the vascular tone. Pulsations of the superficial temporal artery have been shown to diminish in patients with intractable migraine after intracarotid injections of serotonin. It is hypothesized, therefore, that in biochemical terms, the migraine attack may be directly related to a reduction in endogenous serotonin in subjects in whom serotonin metabolism is unstable. Further research is needed, however, to elucidate this biochemical interaction.

4.3.4. Parkinson's Disease

Kinetic disorders affecting the basal ganglia have lately been the subject of many studies, with particular focus on the apparent decrease of dopamine found in these conditions. Reduction of brain dopamine has been found in Parkinson's disease, familial tremor, choreoathetosis, and Huntington's chorea, among others, although reports are still in conflict concerning Huntington's disease. This final section will examine the effects of dopamine in the modification of motor function in Parkinson's disease. Parkinsonism is a disorder of the central nervous system characterized by slowness and poverty of movement, weakness, muscular rigidity, and tremor. The syndrome may be produced by a variety of agents, but in the majority of cases its etiology is unknown. Parkinsonism is classified into three categories: (1) postencephalitic, occurring in a large percentage of cases after epidemic encephalitis; (2) arteriosclerotic, usually affecting patients after age 65; (3) idiopathic, occurring in middle age and more infrequently in elderly patients.

Hornykiewicz (1963) investigated the postmortem content of nor-adrenaline and dopamine in normal and parkinsonian patients and found a marked reduction of dopamine and its metabolite homovanillic acid (HVA) in the substantia nigra of those with idiopathic Parkinson's disease. The most consistent neuroanatomical sign found in postencephalitic parkinsonism is degeneration of the melanin-containing nerve cells in the pars compacta of the substantia nigra. This cellular degeneration is also seen in idiopathic parkinsonism but is pathologically milder. Ehringer and Hornykiewicz (1960) found a good correlation between cell loss in the pars compacta and a deficiency of dopamine and homovanillic acid in post-encephalitic parkinsonism. These results were confirmed in a more elaborate study by Bernheimer et al. (1963). A noradrenaline decrease in the hypo-thalamus and a slight loss of serotonin in various brain regions have also been found in postencephalitic parkinsonism. The reduction in these two monoamines, however, is not as severe as the dopamine loss. Fluorescent microscopic evidence suggests that fiber bundles run from the substantia nigra through the ventral part of the crus cerebri, following the retro-lenticular part of the internal capsule to enter the striatum by way of the internal capsule. However, studies on monkeys show that nigrostriatal fibers probably do not enter the crus cerebri but instead rise from the dorso-medial edge of the substantia nigra and run into the mammillary body, where they follow a lateral course through the subthalamus and internal capsule to enter the striatum. Since the histochemical fluorescence technique does not show any nigrostriatal dopamine fibers in the subthalamus of the rat, the results conflict at least specieswise with other neuroanatomical findings. Bertler (1964) has reported that the destruction of cortical and thalamic regions in the rat does not change the dopamine fluorescent picture in the substantia nigra. Barolin et al. (1964) have described a case of hemiparkin-sonism, in which the concentration of dopamine in the neostriatum was considerably lower in the area contralateral to the side of the tremor while the serotonin content was normal throughout.

Working on the premise that dopamine deficiency in patients with parkinsonism may be a causal factor in the disease, some investigators have attempted to raise the brain dopamine concentration by the administration of its precursor dopa (3,4-dihydroxyphenylalanine). Of the two chemical configurations, L-dopa but not D-dopa has been found beneficial in allaying the severe rigidity encountered in parkinsonian patients, but the effect is only short-lived. D-Dopa, given in concentrations equal to those of L-dopa, has no effects. Reserpine administration is known to induce in animals a form of parkinsonism that can be overcome by the administration of L-dopa. The monoamine cerebral releaser α-methyldopa, which inhibits dopa decarboxylase (probably in endogenous brain tissue), also has a deleterious effect on animals, but unlike reserpine its effects are not counteracted by simultaneous administration of L-dopa.

Rinne *et al.* (1966) studied the excretion of homovanillic acid and vanillylmandelic acid in patients with extrapyramidal disorders, including chorea, athetosis, and cerebral palsy. They found that the mean excretion rate of homovanillic acid in the experimental group was significantly lower than in normal control subjects. The excretion of vanillylmandelic acid was about equal in both groups. It is interesting to note that chorea presents symptoms that are almost opposite to those seen in parkinsonism and actually appears to benefit from reserpine therapy, which causes a depletion of stored dopamine. Other data indicating a nigrostriatal role for catecholamines have come from studies in which lesions were made in the brain stem. Poirier and Sourkes (1965) made lesions in the tegmentum of the pons and midbrain in 19 monkeys and after 9 months noted the following changes: In six animals with unilateral lesions in the ventromedial tegmental area, a loss of cells in the pars compacta ipsilateral to the substantia nigra was evident. Another lesion located elsewhere in the tegmental region of the pons and midbrain did not change the cellular picture of the substantia nigra. In the six animals showing one-sided cell loss, the corresponding caudate nucleus and putamen had decreases of dopamine and noradrenaline content greater than in the nonlesioned side. No catecholamine changes were seen in the remaining 12 animals, in which the nigral cells had remained intact. In a similar experiment, Petsche (1966) made unilateral stereotactic lesions in the rabbit pallidum and found a 10–100 % reduction in the dopamine concentration ipsilateral to the lesion. In lesions involving adjacent structure to the pallidum but excluding the pallidum itself, a similar reduction of dopamine was found in such areas as Forel's field, the internal capsule, and the regio innominata.

Evidence indicating the importance of dopamine in the control of motor functions, including that obtained by study of basal ganglia disorders, *viz.*, Parkinson's disease, can be summarized from the above discussion as follows: (1) presence of nigroneostriatal dopamine neurons as revealed by fluorescence histochemistry; (2) presence of high concentrations of dopamine in the normal neostriatum; (3) loss or reduction of dopamine in basal ganglia structures in parkinsonism, as shown by biochemical studies; (4) development of hemiparkinsonian symptoms when contralateral neostriatal dopamine is decreased; (5) production of experimental model parkinsonian symptoms concomitant with dopamine reduction by nigral lesions; (6) reduction in homovanillic acid, a dopa metabolite, in post-encephalitic parkinsonism. A clinical review of parkinsonism by Denny-Brown (1954) suggests that trauma and emotional elements are of little significance as causative factors in this disorder. Cellular demonstration of monoamines using the histochemical fluorescence technique has shown a functional relationship between the large cells in the pars compacta, the substantia nigra, and the caudate nucleus.

In order to help explain the process linking dopamine to Parkinson's disease and the theoretical basis for employing L-dopa to treat this disorder, a brief review of the biochemistry, pathology, and pharmacology involved in parkinsonism is presented:

4.3.4(a). Biochemistry

In man, as in animals, catecholamines, dopamine, and noradrenaline in the central nervous system are not evenly distributed. The concentration of dopamine is particularly high in extrapyramidal regions such as the neostriatum, globus pallidus, and substantia nigra. The enzymes necessary for the conversion of L-dopa to noradrenaline as well as the enzymes responsible for the metabolism of the amino acid precursors are all found in abundant concentrations in the brain. Metabolic enzymes such as monoamine oxidase and catechol-O-methyltransferase are responsible for the dopamine metabolite homovanillic acid and 3-methoxytyramine. In spite of the large amount of dopamine-β-hydroxylase in the neostriatum, the noradrenaline concentration in this region is relatively low. It has further been reported that if catecholamine breakdown is arrested by inhibiting monoamine oxidase with appropriate drugs, then the rate of accumulation of brain dopamine exceeds that of noradrenaline. Also, those areas rich in dooamine will show a higher concentration of dopamine metabolites than noradrenaline metabolites. These findings as well as the visualization by histochemical fluorescence of central catecholamines in presynaptic terminals argue strongly in favor of a role for dopamine in neurohumoral transmission at the level of extrapyramidal structures.

4.3.4(b). Pathology

Anatomically, the locus coeruleus and the substantia nigra, which are rich in melanin-containing cells, are the elected regions for cellular degeneration in Parkinson's disease. Like the catecholamines, melanin is a derivative of phenylalanine. The importance of tyrosine hydroxylase is therefore evident, as it is involved in the metabolism of dopa, which can then produce melanin or dopamine. Dopamine is oxidized to a melanin-like pigment, and both substances are reduced in the basal ganglia in patients with Parkinson's disease. Ehringer and Hornykiewicz (1960) found a parallel relationship between cell loss in the pars compacta of the substantia nigra and a reduction of dopamine and HVA in postencephalitic parkinsonism. These results have been recently confirmed in more detailed studies. The cellular degeneration of melanin-containing nerve cells is seen in both postencephalitic and idiopathic parkinsonism, while Lewy bodies in the cytoplasm of nigral cells are more common in the idiopathic condition. On the other hand, Alzheimer neurofibrillary tangles occur more often in postencephalitic parkinsonism; electron microscopic study of these neurofibrillary tangles reveals an appearance resembling the histopathology described in patients with Alzheimer's disease.

4.3.4(c). Pharmacology

Information on the pharmacology of dopamine in the central nervous system has been increasing almost geometrically, and the reader is directed to the number of excellent reviews on this subject (Barbeau *et al.*, 1961; Carlsson, 1959; McLennan, 1965; Vogt, 1959). Reserpine and benzo-quinolizine derivatives have been used in studies to produce symptoms resembling those of patients with Parkinson's disease. They can be repro-duced in humans or animals. The *Rauwolfia* alkaloids (*viz.,* reserpine) act to block the incorporation of the cytoplasmic monoamines into the labile granule fraction. As a consequence, the monoamine accumulates in the cytoplasm, where it is exposed to oxidative deamination by monoamine oxidase. There results a fall in the levels of the monoamines, including dopamine, in nigroneostriatal regions. This condition, which causes extrapyramidal symptoms, is reversible after the administration of L-dopa. Other drugs such as the butyrophenones or the phenothiazines can also cause extrapyramidal symptoms (akinesia or dyskinesia), and their action is thought to result from their blocking of synaptic receptors for catechol-amines, in particular dopamine.

Nagatsu *et al.* (1964) ended a long controversy about the origin of dopa in the brain. Up to that time, investigators had suggested that dopa was synthesized by tyrosinase from tyrosine outside the central nervous system and carried to the brain by the circulation. It was not until tyrosine hydroxylase was discovered in the brain as well as in sympathetically innervated tissues and adrenal medulla that the controversy came to an end. Tyrosine hydroxylase is the rate-limiting step in catecholamine synthesis and selectively oxidizes L-tyrosine to L-dopa in the presence of oxygen. D-Tyrosine is not affected in this process. In the brain, dopa is subsequently decarboxylated to form dopamine. The enzyme responsible for this conversion is dopa decarboxylase, a nonspecific catalyst which can also decarboxylate several other aromatic amino acids, including 5-hydroxytryptophan, tyrosine, and histidine. About half of this enzyme is found in the supernatant fraction of the brain homogenate, and recent studies have indicated that it is concentrated in the synaptosomes, or pinched-off nerve endings. Dopa decarboxylase requires pyridoxal phosphate as a cofactor for optimum activity. In the neostriatum and brain stem, there exists a very high dopa decarboxylase activity, for example, 100 $\mu g/g/hr$ compazed to 1 $\mu g/g/hr$ for dopamine-β-hydroxylase (Udenfriend and Creveling, 1959). Studies using dopa-H^3 show that the amine concentrates in the nerve endings and unmyelinated axons; it is seen as dense cores on electron microscopy (Fuxe et al., 1965). There are at least two uptake mechanisms involved in dopa metabolism: (1) reuptake at the presynaptic neuronal cell membrane, and (2) transfer of monoamine from cytoplasm to storage site within the same neuron.

Homovanillic acid, the major final metabolite of dopamine, is found localized in those areas of the brain containing a high level of dopamine. HVA is probably transported out of the brain by an active system, and its levels in the spinal fluid have been used by investigators and clinicians as an index of dopamine concentration in the brain. These data (and much more when finally analyzed, classified, and digested) lead to the inevitable conclusion that if one common denominator in Parkinson's disease can be found, it is the lack of dopaminergic modulation. It therefore is a logical assumption that if dopamine deficiency exists in nigroneostriatal structures it merely remains to replenish or increase the levels of brain dopamine by the administration of dopa, its direct precursor, since dopamine itself has been found not to penetrate the blood–brain barrier in any significant amounts. This simple and correct logic was put into use as early as 1961, when investigators first began using L-dopa to treat Parkinson's disease.

Reports of the first clinical trials with L-dopa in Parkinson's disease (Birkmayer and Hornykiewicz, 1961; Gerstenbrand et al., 1963; Umbach and Tzavellas, 1965; Bruno and Brigida, 1965; Barbeau et al., 1962) were received with only mild enthusiasm. The effects of L-dopa, though favorable, were short-lasting. In these studies, the L-dopa was given intravenously and orally in doses never exceeding 500 mg, since animal studies had indicated that the LD_{50} of this amino acid is 580 mg/kg in neonatal rats and 610 mg/kg in adult rabbits, while dogs given a total daily dose of 1 g/kg die or are moribund within 6 weeks and show subendocardial petechial hemorrhages at necropsy. Intravenous administration of L-dopa at 350 mg was seen to increase the blood pressure of patients 50 mm Hg on the average, while doses below 100 mg resulted in a 20 mm Hg decrease in other patients. Side-effects after L-dopa were varied and not encouraging. They frequently included nausea with vomiting, vertigo, profuse sweating, anxiety, cardiac irregularities, anorexia, and dizziness. But in spite of the precautions urged in the use of L-dopa in humans, Cotzias et al. (1967) began treating 16 parkinsonian patients with the more toxic and 50% less active isomer DL-dopa, reaching fantastic dose levels of 16 g per day in some patients. This incredibly bold "damn the torpedos" approach to therapy, which can only be described as exercising more courage than common sense, *did* ameliorate the akinesia and muscular atony in 12 patients in spite of severe but reversible granulocytopenia in four patients and vacuolation of myelotes and myelocytes in another four of the 12 who improved. Films showing the dramatic effects of the dopa treatment in patients suffering from akinesia and dyskinesia and their subsequent improvement in walking ability were promptly released to the lay press, making Cotzias and his colleagues overnight celebrities. These exceedingly high doses, which Cotzias had described as "heroic," were reduced by half in subsequent studies when Cotzias became aware of the increased activity and diminishing toxic effects reported 6 years earlier for the L-form of dopa instead of the DL-isomer. Since success has many friends and caution very few, what followed was a flurry of experimental human trials using L-dopa. These tests confirmed the beneficial effects of the oral L-dopa treatment. Some of these clinical trials were well controlled and observed due caution in treatment with this amine in hig doses, while others left much to be desired. At the time of this writing, L-dopa has been approved by the Food and Drug Administration and is available for prescription use.

Much less noise was made when one of the most exciting biochemical hunts in the search for an effective drug treatment for Parkinson's disease began shortly after a breakthrough in our understanding of blood–brain barrier mechanisms. Bertler *et al.* (1963*c*) in Sweden first reported that the brain barrier to L-dopa could be found in the brain capillaries. In experiments with animals, these capillaries were found to contain dopa decarboxylase, which trapped the L-dopa when the latter was administered and quickly converted the amino acid to dopamine. In addition, the endothelial capillaries were also found to contain monoamine oxidase, which deaminated the accumulated dopamine to its metabolites. One way in which the L-dopa could cross the brain barrier in significant amounts depended on the use of a dopa decarboxylase inhibitor that would inactivate the peripheral brain barrier decarboxylase enzyme. A then newly synthesized compound, Ro 4-4602, did the trick. Histochemical fluorescence of rat brains revealed that after Ro 4-4602 treatment followed by L-dopa the amino acid passed the barrier and diffused throughout the nerve tissue. However, it was found that if the Ro 4-4602 was used in strong doses (100 mg/kg and above), the concentrations of endogenous brain dopamine did not rise significantly after the administration of L-dopa. The reason for this was that at this dose, Ro 4-4602 not only inhibited peripheral dopa decarboxylase but also endogenous dopa decarboxylase, thus limiting the rate of synthesis of dopa to dopamine in such structures as the neostriatum and the substantia nigra. A few years later, at the psychiatric clinic of the University of Geneva, it was shown (Constantinidis *et al.*, 1968; de la Torre, 1968*b*; Constantinidis *et al.*, 1967; 1969*a*) that very small doses of Ro 4-4602 could greatly enhance the penetration of L-dopa through the blood–brain barrier. This phenomenon was demonstrated by using the histochemical fluorescence technique, which visualizes the monoamines *in situ* after various drug treatments. The clear evidence of a blood–brain barrier mechanism for L-dopa is seen when the amino acid is injected into animals previously treated with a MAO inhibitor such as nialamide. Very bright, green fluorescent brain capillaries appear, densely distributed throughout the brain except in two zones, the infundibulum and the area postrema, both of which are outside the blood–brain barrier. The fluorescent capillaries in the brain appear to be evenly distributed except in two hypo-thalamic nuclei, the paraventricular nucleus (pars magnocellularis and pars parvocellularis) and the supraoptic nucleus, both of which contain a more dense population of capillaries. The capillary fluorescence after nialamide plus L-dopa is due to the rapid decarboxylation of this amino acid by peripheral dopa decarboxylase to dopamine, and since no attack by monoamine oxidase is possible, due to its inhibition by nialamide, the accumulated dopamine remains trapped in the capillaries unable to pene-trate into the brain parenchyma (see Section 3.3.4). In contrast, nialamide pretreatment followed by small doses of Ro 4-4602 (2–50 mg/kg) allows sub-sequently administered L-dopa to penetrate into discrete regions of the brain, particularly the hypothalamus. This penetration is seen as a green diffuse area in the nerve tissue, devoid of any background capillary fluorescence.

As a result of such experiments, the following conclusions have been reached: (1) progressive dose increases of Ro 4-4602 indicate that the blood–brain barrier is weakest to dopa in the anteromedial region of the hypothalamus, which includes the ventromedial and suprachiasmatic nuclei; (2) the barrier system appears to be strongest to L-dopa in the superomedial and lateral aspects of the hypothalamus and moderately active in the preoptic and lateral hypothalamic nuclei; (3) dopa penetration after dopa decarboxylase inhibition appears to be related to vascular density—the more richly vascularized the region, the more difficult it becomes for dopa to enter the brain parenchyma; (4) the concentration of dopamine in nigroneostriatal structures is markedly enhanced after peripheral decarboxylase inhibition compared to that after L-dopa injec-tions only; (5) the severity of autonomic signs that occur after the use of MAO inhibitors and L-dopa is diminished or completely abolished fol-lowing peripheral decarboxylase inhibition; (see Table 14) (6) almost total penetration of L-dopa is achieved after only 50 mg/kg of Ro 4-4602 in all regions of the brain; (7) the use of a peripheral decarboxylase inhibitor prior to L-dopa treatment could be of extreme importance in potentiating the effects of this amino acid in basal ganglia disorders, particularly Parkinson's disease.

These findings led Tissot *et al.* (1969) to treat 25 subjects suffering from idiopathic and postencephalitic parkinsonism with Ro 4-4602 and L-dopa. The duration of the disorder in these patients ranged from 1 to 23 years with a mean of $7\frac{1}{2}$ years. Both drugs were administered by mouth in doses of between 450–1200 mg for Ro 4-4602 and 450–900 mg for L-dopa. The effects of the drug combination were dramatic. Of the 21 patients diagnosed as having classical Parkinson's disease (four patients had senile tremors), very favorable results were seen in 18 patients, including three apparent cures; in two there were no effects, and in one a serious hemolytic anemia occurred. Eleven patients had a mild and reversible form of hemolytic anemia, which disappeared spontaneously without reduction in the dose. The usual side effects seen after dopa treatment—nausea, vomiting, and autonomic disturbances—were rare and could be avoided with adequate dosing. The amelioration of the extrapyramidal signs with maximum improvement was seen on the average after a few days, compared to that reported with use of L-dopa alone of rarely less than 6 weeks and usually many months in patients receiving 5–8 g per day of the amine. Siegfried *et al.* (1969), comparing the effect of L-dopa alone with that of L-dopa combined with a peripheral decarboxylase inhibitor, concluded that the drug combination had been shown objectively to be more effective than L-dopa alone. These reports were variously confirmed by Barbeau and Gillo-Joffroy (1969) and by Birkmayer (1969) and recently in this country by Yahr *et al.* (1972).

By itself, L-dopa may bring relief of akinesia in humans when doses in excess of 4 g per day are used. The logic behind this treatment is that by saturating the peripheral enzyme dopa decarboxylase, some of the administered L-dopa will leak through the capillary endothelial cells into the brain neuronal tissue to enhance dopamine concentrations there. However, the consequences of using high doses of L-dopa can become a serious disadvantage even when relief of some of the symptoms is evident. In humans, 1200 mg per day of the amino acid can cause dizziness, tinnitus, weakness, nausea, and vomiting (Pazzagli and Amaducci, 1966). It has also been reported to produce transient thrombocytopenia, confusion, and hallucinations (Steg, 1969). In a recent study using L-dopa on 100 patients with Parkinson's disease, Duvoisin *et al.* (1969) reported that during a 19-month period symptoms such as cardiac arrhythmias, postural hypotension, and involuntary movements were the major dose-limiting side effects, observed in over three-fourths of the patients, and that while leukopenia was transient and rare, some minor elevations of SGOT and serum urea nitrogen were seen. While at the present it appears that L-dopa is the most effective drug in the treatment of parkinsonism, it was the conclusion of Duvoisin's group that the numerous side effects of L-dopa alone complicate its use and often limit its benefits.

It is still not yet known what the long-term effects of such high doses of a probable neurotransmitter precursor may have on the motor, sensory, and psychic systems in humans. It is also obvious that one way of avoiding gross levels of systemic L-dopa after its administration is to reduce the dose drastically without reducing its potency in extrapyramidal structures. It appears, therefore, that the logical answer may lie in the enzymatic inhibition of dopa decarboxylase at the capillary level with a subsequent reduced dose of L-dopa. What must be found now is an active, nontoxic decarboxylase inhibitor that can be used safely in long-term therapy for the management and possible control of the degenerative, progressive course of Parkinson's disease. Ideally, such a combined form of therapy would correct the existing symptoms as well as the potential pathology of the disorder and would prevent actual symptoms or arrest the progression of chronic symptoms. This in reality is what successful drug therapy should resolve.

In the preceding sections, we have examined only superficially the known and potential physiological effects of manipulating monoamines in the brain. The psychopharmacology of drugs affecting the mind and body is only in its childhood, but it undoubtedly will have far-reaching consequences in the near future as our knowledge of the function of the nervous system increases. How the applications derived from these early experiments will affect human behavior and society as a whole is difficult to predict. Neither should the responsibility of scientists involved in this field be overlooked. This is not a warning to indulge in new morality or ethics—it is merely a recognition that within our reach is the power to tamper with nature's controls. As the application of this knowledge becomes available to society, a technology in the service of human values will be hopefully created.

Appendix :

Abbreviations Used in Text

A—adrenaline
ACh—acetylcholine
AChase—acetylcholinesterase
ACTH—adrenocorticotropic hormone
ADH—antidiuretic hormone
ATP—adenosine triphosphate
ATPase—adenosine triphosphatase
BOL-148—2-brom-lysergic acid diethylamide
CA—catecholamine
CD—chlordiazepoxide
CNS—central nervous system
COMT—catechol-O-methyltransferase
CPZ—chlorpromazine
DA—dopamine
DCI—peripheral dopa decarboxylase inhibitor
DET—diethyltryptamine
DHMA—4-dihydroxymandelic acid
DHPG—3,4-dihydroxyphenylglycol
DIMPEA—3,4-dimethoxyphenylethylamine
DMI—desmethylimipramine
DMSO—dimethylsulfoxide
DMT—dimethyltryptamine
Dopa—3,4-dihydroxyphenylalanine
Dopamine—3,4-dihydroxyphenylethylamine
DOPS—dihydroxyphenylserine
DPH—diphenylhydantoin
ECS—electroconvulsive shock
EEG—electroencephalogram
FLA-63—isomer of tetraethylthiouram disulfide

5-HIAA—5-hydroxyindoleacetic acid
HIOMT—hydroxy-O-methyltransferase
HMMA—4-hydroxy-3-methoxymandelic acid
HMPG—4-hydroxy-3-methoxyphenylglycol
5-HT—5-hydroxytryptamine (serotonin)
5-HTP—5-hydroxytryptophan
HVA—homovanillic acid
LHRF—luteinizing hormone releasing factor
LSD—lysergic acid diethylamide
MA—monoamine
MAO—monoamine oxidase
MFB—medial forebrain bundle
MK-485—α-methyldopa hydrazine
MK-486—L-isomer of MK-485
MO-911—pargyline
NA—noradrenaline
NK—neuroketone
NSD-1034—N-(3-hydroxybenzyl)-N'-methylhydrazine
6-OHDA—6-hydroxydopamine
p-CPA—p-chlorophenylalanine
PGO—ponto-geniculo-occipital spikes
PIH—phenylisopropylhydrazine
PKU—phenylketonuria
PS—paradoxical sleep
P/S ratio—particulate/supernatant ratio
Ro 4-4602—[N-(DL-seryl)-N'-2,3,4-trihydroxybenzyl] hydrazine
SGOT—serum glutamic oxalacetic transaminase
SS—slow wave sleep
U-14,624—1-phenyl-3-(2-thiazolyl)-2-thiourea
WIN 18501-2—1-(5,6-dimethoxy-2-methyl-3-indole)-ethyl-4-phenyl-
    piperazine

References

Adams, D. P. (1968). Cells related to fighting behavior from midbrain central gray neuropil of the cat. *Science* **159**:894–896.

Adams-Ray, J., Dahlström, A., Fuxe, K., and Hillarp, N. A. (1964). Mast cells and monoamines. *Experientia* **20**:80–82.

Aghajanian, G.K., and Bloom, F.E. (1966). Electron-miscroscopic autoradiography of rat hypothalamus after intraventricular H-3 norepinephrine. *Science* **153**:308–310.

Aghajanian, G. K., Rosencrans, J. A., and Sheard, M. H. (1967). Serotonin release in the forebrain by stimulation of the midbrain raphé. *Science* **156**:402–403.

Aghajanian, G. K., Fotte, W. E., and Sheard, M. H. (1970). Action of psychotogenic drugs on single midbrain raphé neurons. *J. Pharmacol. Exptl. Therap.* **171**:178–187.

Ajuriaguerra, J. de (1961). Monoamines et structures anatomofonctionelles cérébrales. In *Monoamines et système nerveux central,* Symposium Bel-Air, Georg & Cie, Geneva, pp. 277–293.

Akamayev, E. G. (1965). Catecholamines of the zona palisadica of the eminentia mediana of the hypothalamus in adrenalectomy, administration of hydrocortisone and stress. *Z. Mikr. Anat. Forsch.* **74**:83–91.

Akimov, N. E. (1937). The effect of adrenalin upon maze learning in white rats. *Psychol. Abst.* **11**:123.

Albe-Fessard, D., Stutinsky, F., and Libouban, S. (1966). *Atlas stéreotaxique du diencéphale du rat blanc,* Centre National de la Recherche Scientifique, Paris.

Alpers, B. J. (1937). Relation of the hypothalamus to disorders of the personality. *Arch. Neurol. Psychiat.* **38**:291.

Amin, A. H., Crawford, T. B. B., and Gaddum, J. H. (1954). The distribution of substance P and 5-hydroxytryptamine in the central nervous system of the dog. *J. Physiol. (Lond.)* **126**:596.

Andén, N. E. (1964). On the mechanism of noradrenaline depletion by α-methylmetatyrosine and metaraminol. *Acta Pharmacol.* **21**:260–271.

Andén, N. E. (1968). Discussion of serotonin and dopamine in the extrapyramidal system. *Adv. Pharmacol.* **6A**:347–349.

Andén, N. E., Lundberg, A., Rosengren, E., and Vyklicky, L. (1963). The effect of dopa on spinal reflexes from the FRA (flexor reflex afferents). *Experientia* **19**:654–655.

Andén, N. E., Carlsson, A., Dahlström, A., Fuxe, K., Hillarp, N. A., and Larsson, K. (1964a). Demonstration and mapping out of nigro-neostriatal dopamine neurons. *Life Sci.* **3**:523–530.

Andén, N. E., Carlsson, A., Hillarp, N. A., and Magnusson, T. (1964b). 5-Hydroxytryptamine release by nerve stimulation of the spinal cord. *Life Sci.* **3**:473–478.

Andén, N. E., Jukes, M. G. M., and Lundberg, A. (1964c). Spinal reflexes and monoamine liberation. *Nature* **202**:1222–1223.

Andén, N. E., Dahlström, A., Fuxe, K., and Larsson, K. (1965). Further evidence for the presence of nigro-neostriatal dopamine neurons in the rat. *Am. J. Anat.* **116**:329–334.

Andén, N. E., Dahlström, A., Fuxe, K., and Larsson, K. (1966a). Functional role of the nigro-neostriatal dopamine neurons. *Acta Pharmacol.* **24**:263–274.

Andén, N. E., Dahlström, A., Fuxe, K., and Larsson, K. (1966b). Ascending noradrenaline neurons from the pons and the medulla oblongata. *Experientia* **22**:44–45.

Andén, N. E., Corrodi, H., Fuxe, K., and Hökfelt, T. (1968). Evidence for a central 5-hydroxytryptamine receptor stimulation by lysergic acid diethalamide. *Brit. J. Pharmacol.* **34**:1.

Angel, C., and Burkett, M. L. (1966). Adrenalectomy, stress and the blood–brain barrier. *Dis. nerv. syst.* **27**:389–393.

Antonelli, A. R., Bertaccini, G., and Mantegazzini, P. (1961). Relationship between mesencephalic–hypothalamic concentration of 5-hydroxytryptamine and cortical electrical activity of cats. *J. Neurochem.* **8**:157–158.

Antunas-Rodriguez, J. (1963). Hypothalamic control of NaCl and water intake. *Acta Physiol. Lat.-Am.* **13**:94–100.

Aprison, M. H., Honson, K. M., and Austin, D. C. (1959). Studies on serum oxidase (ceruloplasmin) inhibition of tryptophan metabolites. *J. Nerv. Ment. Dis.* **128**:249–255.

Aprison, M. H., Wolf, M. A., and Poulos, G. L. (1962). Neurochemical correlates of behavior. III. Variation of serotonin content in several brain areas and peripheral tissues of the pigeon following 5-hydroxytryptophan administration. *J. Neurochem.* **9**:575–84.

Arees, E. A., and Mayer, J. (1967). Anatomical connections between medial and lateral regions of the hypothalamus concerned with food intake. *Science* **157**:1574–1575.

Arioka, I., and Tanimukas, H. (1957). Histochemical studies on monoamine oxidase in the midbrain of the mouse. *J. Neurochem.* **1**:311–315.

Armstrong, M. D., and McMillan, A. (1957). Identification of a major urinary metabolite of norepinephrine. *Fed. Proc.* **16**:146.

Armstrong, M. D., McMillan, A., and Shaw, K. N. (1957). 3-Methoxy-4-hydroxy-mandelic acid, a urinary metabolite of norepinephrine. *Biochim. Biophys. Acta* **25**:422–423.

Aserinsky, E., and Kleitman, N. (1953). Regularly occurring periods of eye motility and concomitant phenomena during sleep. *Science* **118**:273.

Ax, A. F. (1953). The physiological differentiation between fear and anger in humans. *Psychosomat. Med.* **15**:433.

Axelrod, J. (1957). O-methylation of epinephrine and other catechols *in vitro* and *in vivo*. *Science* **126**:400.

Axelrod, J. (1962). The enzymatic N-methylation of serotonin and other amines. *J. Pharmacol. Exptl. Therap.* **138**:28–33.

Axelrod, J., and Tomchick, R. (1958). Enzymatic O-methylation of epinephrine and other catechols. *J. Biol. Chem.* **233**:702–705.

Axelrod, J., and Weissback, H. (1960). Enzymatic O-methylation of N-acetyl-serotonin to melatonin. *Science* **131**:132.

Azcoaga, J. E. (1965). Senility of the neurons of the large cell nuclei of the hypothalamus. *Arch. Histol.* **9**:40–48.

Back, M. S., Gosselin, L., Dresse, A., and Renseon, J. (1959). Inhibition of O-methyltransferase by catechol and sensitation to epinephrine. *Science* **130**:453–454.

Baechtold, H. P., and Pletscher, A. (1957). Einfluss von Isonicotinsäure-Hydraziden auf den Verlauf der Körpertemperatur nach Reserpin Monoaminen und Chlorpromazin. *Experientia* **13**:163.

Baird, J. P. C., and Lewis, J. J. (1964). The effects of cocaine, amphetamine and some amphetamine-like compounds on the *in vivo* levels of noradrenaline and dopamine in the rat brain. *Biochem. Pharmacol.* **13**:1475–1482.

Baldessarini, R. J., and Kopin, I. J. (1966). Tritiated norepinephrine: Release from brain slices by electrical stimulation. *Science* **152**: 1630–1631.

Baldridge, R. C., Borofsky, L., Baird, III, H., Reichle, F., and Bullock, D. (1959). Relationship of serum phenylalamine levels and ability of phenylketonurics to hydroxylate tryptophan. *Proc. Soc. Exp. Biol. Med.* **100**: 529.

Banerjee, S., and Agarwal, P. S. (1968). Tryptophan–nicotinic acid metabolism in schizophrenia. *Proc. Soc. Exptl. Biol.* **97**: 657–659.

Barbeau, A. (1960). Preliminary observations on abnormal catecholamine metabolism in basal ganglia diseases. *Neurology* **10**: 446–451.

Barbeau, A. (1963). Etudes récentes sur les catécholamines. *Un. med. Can.* **92**: 42–51.

Barbeau, A. (1967). The "pink spot," 3,4-dimethoxyphenylethylamine and dopamine: Relationship to Parkinson's disease and to schizophrenia. *Rev. Can. Biol.* **26**: 55–77.

Barbeau, A. (1970). Dopamine and central nervous system control of blood pressure. *Fifth Canad. Congr. of Neurolog. Sci.*, Toronto, June 10–13.

Barbeau, A., and Gillo-Joffroy, L. (1969). Treatment of Parkinson's disease with L-dopa and Ro 4-4602. Ninth Intern. Congr. Neurol., New York, Sept. 20–27.

Barbeau, A., and Singh, P. (1965). Effect of 3,4-dimethoxyphenylethylamine injections on the concentration of catecholamines in the rat brain. *Rev. Can. Biol.* **24**: 229–232.

Barbeau, A., and Sourkes, T. L. (1961). Some biochemical aspects of extrapyramidal diseases. In Bordeleau, J. M. (ed.), *Extrapyramidal System and Neuroleptics,* Montreal, pp. 101–107.

Barbeau, A., Murphy, G. F., and Sourkes, T. L. (1961). Excretion of dopamine in diseases of basal ganglia. *Science* **133**: 1706–1707.

Barbeau, A., Sourkes, T. L., and Murphy, G. F. (1962). Les catecholamines dans la maladie de Parkinson. In Ajuriaguerra, J. de (ed.), *Monoamines et système nerveux central,* Georg & Cie, Geneva, pp. 247–262.

Barchas, J. D., and Freedman, D. X. (1963). Brain amines: Response to physiological stress. *Biochem. Pharmacol.* **12**: 1232–1235.

Bard, P. (1928). A diencephalic mechanism for the expression of rage with special reference to the sympathetic nervous system. *Am. J. Physiol.* **84**: 490.

Bargemann, W. (1954). *Das Zwischenhirn-Hypophysensystem,* Julius Springer, Berlin.

Barolin, G. S., Bernheimer, H., and Hornykiewicz, O. (1964). Difference in the behavior of dopamine (3-hydroxytyramine) in each side of the brain in a case of hemiparkinsonism. *Schweiz. Arch. Neurol. Psychiat.* **941**: 241–248.

Barraclough, C. A. (1956). Blockade of the release of pituitary gonadotrophin by reserpine. *Fed. Proc.* **14**: 9.

Barry, J., and Leonardelli, J. (1967). Etude de la topographie des fibres et des neurons monoaminergiques au niveau de l'hypothalamus chez le cobaye normal et stereotaxé. *Compt. Rend. Acad. Sci. (Paris)* **265**: 557–560.

Bartholini, G., and Pletscher, A. (1968). Cerebral accumulation and metabolism of C^{14} dopa after selective inhibition of peripheral decarboxylase. *J. Pharmacol. Exptl. Therap.* **16**: 14–20.

Bartholini, G., Richards, J. G., and Pletscher, A. (1970). Dissociation between biochemical and ultrastructural effects of 6-hydroxydopamine in rat brain. *Experientia* **26**: 143.

Bartlet, A. L. (1960). The 5-HT content of mouse brain and whole mice treatment with some drugs affecting the C.N.S. *Brit. J. Pharmacol.* **5**: 140–146.

Bartlet, A. L. (1965). The influence of chlorpromazine on the metabolism of 5-hydroxytryptamine in the mouse. *Brit. J. Pharmacol.* **24**: 497–509.

Bartonicék, V. (1965a). The influence of high doses of imipramine on the intraneuronal levels of brain monoamines in albino rats. *Activ. Nerv. Sup.* **7**: 279.

Bartonicék, V. (1965b). Brain monoamines in the specific neurons of albino rats as influenced by tetrabenazine. *Activ. Nerv. Sup.* **7**: 264.

Bartonicék, V. (1965c). Failure of imipramine to influence serotonin and catecholamine levels in specific neurons of rat brain. *Med. Exp.* **12**: 395–398.

Bartonicék, V. (1965d). Increase of serotonin levels in the specific serotonergic neurons of rat brain caused by phenelzine. *Med. exp.* **13**:184–188.

Bartonicék, V., Dahlström, A., and Fuxe K. (1964). Effects of certain psychopharmaca on intraneuronal levels of 5-HT and catecholamines in the specific monoamine neurons of the rat brain. *Experientia* **20**:690–691.

Bazelon, M., Paine, R. S., Cowie, V. A., Hunt, P., Houck, J. C., and Mahanand, D. (1967). Reversal of hypotonia in infants with Down's syndrome by administration of 5-hydroxytryptophan. *Lancet* **1**:1130–1133.

Bell, C. E., and Sommerville, A. P. (1966). A new fluorescence method for the detection and possible quantitative assay of some catecholamine and tryptamine derivatives on paper. *Biochem. J.* **98**:1C–3C.

Belleau, B., and Moran, J. (1963). Deuterium isotope effects in relation to the chemical mechanism of monoamine oxidase. *Ann. N. Y. Acad. Sci.* **107**:822–839.

Benitez, H. H., Murray, M. R., and Woolley, D. W. (1955). Effects of serotonin and certain of its antagonists upon oligodendroglial cells *in vitro*. Second International Congress of Neuropathology, London, September 12–17, 1955.

Bennet, D. S., and Giarman, N. J. (1965). Schedule of appearance of 5-hydroxytryptamine (serotonin) and associated enzymes in the developing rat brain. *J. Neurochem.* **12**:911–918.

Berger, J. (1962). Phenylpyruvic idiocy and tyrosinosis. *Bull. Schweiz. Akad. Med. Wiss.* **17**:334.

Berger, F. M., Campbell, G. L., Hendley, C. D., Ludwig, B. J., and Lynes, T. E. (1957). The action of tranquilizers on brain potentials and serotonin. *Ann. N. Y. Acad. Sci.* **66**:686–694.

Bergsman, A. (1959). The urinary excretion of adrenaline and noradrenaline in some mental diseases: A clinical and experimental study. *Acta Psychiat. Scand.* **34**: (suppl. 133) 1–107.

Bernardis, L. L. (1963). Food intake patterns from weaning to adulthood in male and female rats with hypothalamic lesion. *Experientia* **19**:541–543.

Bernardis, L. L., and Montemurro, D. G. (1963). The response of male and female rats with hypothalamic lesions to low and high environmental temperatures. *Experientia* **19**:26–27.

Bernheimer, H. (1964). Distribution of homovanillic acid in the human brain *Nature (Lond.)* **204**:587–588.

Bernheimer, H. (1965). On the occurrence of 5-hydroxyindoleacetic acid in the human brain. *Klin. Wschr.* **43**:1119.

Bernheimer, H., Birkmayer, W., and Hornykiewicz, O. (1961). Verteilung des 5-Hydroxytryptamins (Serotonin im Gehirn des Menschen und sein Verhalten bei Patienten mit Parkinson-Syndrom). *Klin. Wschr.* **39**:1056–1059.

Bernheimer, H., Birkmayer, W., and Hornykiewicz, O. (1963). Zur Biochemie des Parkinson-Syndroms des Menschen, Einfluss der Monoaminoxydase-Hemmer-Therapie auf die Konzentration des Dopamins, Noradrenalins und 5-Hydroxytryptamins im Gehirn. *Klin. Wschr.* **41**:465.

Bertaccini, G. (1959). Effect of convulsant treatment on the 5-HT content of brain and other tissues of the rat. *J. Neurochem.* **4**:217–222.

Bertaccini, G. (1960). Tissue 5-hydroxytryptamine and urinary 5-hydroxyindoleacetic acid after partial or total removal of the gastrointestinal tract in the rat. *J. Physiol. (Lond.)* **153**:239.

Bertler, A. (1961a). Effect of reserpine on the storage of catecholamines in brain and other tissues. *Acta Physiol. Scand.* **51**:75–83.

Bertler, A. (1961b). Occurrence and localization of catecholamines in the human brain. *Acta Physiol. Scand.* **51**:97–107.

Bertler, A. (1964). *Biochemical and Neurophysiological Correlations of Centrally Acting Drugs*. Pergamon Press, Oxford.

Bertler, A., and Rosengren, E. (1959a). Occurrence and distribution of dopamine in brain and other tissues. *Experientia* **15**:10–11.

Bertler, A., and Rosengren, E. (1959b). On the distribution of monoamines and of enzymes responsible for their formation. *Experientia* **15**:382–384.

Bertler, A., Carlsson, A., and Rosengren, E. (1963a). Fluorometric method for differential estimation of the 3-O-methylated derivatives of adrenaline and noradrenaline (metanephrine and normetanephrine). *Clin. Chim. Acta* **4**:456–457.

Bertler, A., Falck, B., and Owman, C. (1963b). Cellular localization of 5-hydroxytryptamine in the rat pineal gland. *Kgl. Fisiogr. Sallsk. Hdl.* **33** (2):13–16.

Bertler, A., Falck, B., and Rosengren, E. (1963c). The direct demonstration of a barrier mechanism in the brain capillaries. *Acta Pharmacol.* **20**:317–321.

Bertler, A., Falck, B., Owman, C., and Rosengren, E. (1966). The localization of monoaminergic blood-brain barrier mechanisms. *Pharmacol. Rev.* **18**:369–385.

Bertrand, G., Jasper, H., and Wong, A. (1967). Microelectrode study of the thalamus in patients with dyskinesias. Third International Symposium of Stereoencephalotomy, Madrid, Spain.

Besendorf, H., and Pletscher, A. (1956). Beeinfluss zentraler Wirkungen von Reserpin und 5-Hydroxytryptamine durch Isonicotinsaurehydrazide. *Helv. Physiol. Pharmacol.* Acta **14**:383–398.

Bianchi, C. (1957). Reserpine and serotonin in experimental convulsions. *Nature* **179**:202–203.

Biel, J. H., Horita, A., and Drukker, A. E. (1964). Monoamine oxidase inhibitors (hydrazines). In Gorden, M. (ed.), *Psychopharmacological Agents,* Academic Press, New York, Vol. 1, pp. 349–443.

Birkmayer, W. (1969). Experimentelle Ergebnisse über die Kombinationsbehandlung des Parkinson-Syndroms mit L-Dopa und einem Decarboxylase-Hemmer Ro 4–4602. *Wien. Klin. Wschr.* **81**:677.

Birkmayer, W., and Hornykiewicz, O. (1961). Der L-Dioxyphenylalain (Dopa)-Effekt bei der Parkinson-Akinese. *Wien Klin. Wschr.* **73**:787.

Birkmayer, W., and Mentasti, M. (1967). Weitere experimentelle Untersuchungen über den Catecholaminstoffwechsel bei extrapyramidalen Ekrankungen (Parkinson- und Chorea-Syndrom) *Arch. Psychiat. Zeitschrift f. d. ges. Neurol.* **210**:29.

Björklund, A., Falck, B., Hromek, F., and Owman, C. (1969). An enzymic mechanism for monoamine precursors in the newly-forming brain capillaries following electrolytic or mechanical lesions. *J. Neurochem.* **16**:1605–1608.

Björklund, A., Falck, B., Hromek, F., Owman, C., and West, K. A. (1970). Identification and terminal distribution of the tubero-hypophyseal monoamine fibre systems in the rat by means of stereotaxic and microspectrofluorimetric techniques. *Brain Res.* **17**:1–23.

Blaschko, H. (1939). The specific action of L-dopa decarboxylase. *J. Physiol. (Lond.)* **96**:50P–51P.

Blaschko, H. (1952). Amine oxidase and amine metabolism. *Pharmacol. Rev.* **4**:415–458.

Bogdanski, D. F., and Brodie, B. B. (1969). The effects of inorganic ions on the storage and uptake of H^3-norepinephrine by rat heart slices. *J. Pharmacol. Exptl. Therap.* **165**:181 189.

Bogdanski, D. F., Pletscher, A., Brodie, B. B., and Udenfriend, S. (1956). Identification and assay of serotonin in brain. *J. Pharmacol. Exptl. Therap.* **117**:82.

Bonnycastle, D. D., Giarman, N. J., and Paasonen, M. D. (1957). Anticonvulsant compounds and 5-hydroxytryptamine in rats' brain. *Brit. J. Pharmacol.* **12**:228.

Bonvallet, M. (1965). Hyperactivity induced by amphetamine in mice during forced exercise and noradrenaline and 3,4-hydroxytyramine (dopamine) content of brain and heart *J. Physiol. (Paris)* **57**:551–552.

Booth, D. A. (1967). Localization of the adrenergic feeding system in the rat diencephalon. *Science* **158**:515–517.

Boullin, D. J., and O'Brien, R. A. (1970). Accumulation of dopamine by blood platelets from normal subjects and Parkinsonian patients under treatment with L-dopa. *Brit. J. Pharmacol.* **39**:779–788.

Boulton, A. A., Pollitt, R. J., and Major, J. R. (1967). Identity of a urinary "pink spot" in schizophrenia and Parkinson's disease. *Nature (Lond.)* **215**:132–134.

Bourdillon, R. E., and Ridges, A. P. (1967). 3, 4-Dimethoxyphenylethylamine in schizophrenia. In Himwich, H. E., Kety, S., and Smythies, J. R. (eds.), *Amines and Schizophrenia*, Pergamon Press, Oxford, pp. 43–50.

Bourne, B. B. (1965). Metabolism of amines in the chick during embryonic development. *Life Sci.* **4**:583–591.

Bradley, P. B., and Woltencroft, J. H. (1965). Actions of drugs on single neurons in the brainstem. *Brit. Med. Bull.* **21**:15–18.

Brady, J. V. (1957). Assessment of drugs on emotional behavior. *Science* **123**:1033–1034.

Brendel, W., and Usinger, W. (1961). Die Bedeutung der Hirntemperatur für die Auslösung Kältezitterns. Ein Beitrag zur Frage der cerebralen Kältereception. *Pflügers Arch. Ges. Physiol.* **274**:77–78.

Brengelmann, J. C., Pare, C. M. B., and Sandler, M. (1959). Effect of 5-HTP on schizophrenia. *J. Ment. Sci.* **55**:770–776.

Brink, J. J., and Stein, D. G. (1967). Pemoline levels in brain: Enhancement by dimethyl sulfoxide. *Science* **158**:1479–1480.

Brodal, A. (1970). *Neurological Anatomy in Relation to Clinical Medicine*, Oxford University Press, New York.

Brodal, A., Taber, E., and Walberg, F. (1960). The raphé nuclei of the brain stem in the cat. *J. Comp. Neurol.* **114**:239–259.

Brodie, B. B., and Beaven, M. A. (1963). Neurochemical transducer systems. *Med. Exp.* **8**: 320.

Brodie, B. B., and Costa, E. (1962). Some current views on brain monoamines. In Ajuriaguerra, J. de (ed.), *Monoamines et système nerveux central*, Symposium Bel-Air, Georg & Cie, Geneva. pp. 13–49.

Brodie, B. B., and Shore, P. A. (1957). A concept of the role of serotonin and norepinephrine as chemical mediators in the brain. *Ann. N. Y. Acad. Sci.* **66**:631–642.

Brodie, B. B., Pletscher, A., and Shore, P. A. (1955). Evidence that serotonin has a role in brain function. *Science* **122**:968.

Brodie, B. B., Pletscher, A., and Shore, P. A. (1956a). Possible role of serotonin in brain function and in reserpine action. *J. Pharmacol.* **116**:9.

Brodie, B. B., Shore, P. A., and Pletscher, A. (1956b). Serotonin releasing activity limited to rauwolfia alkaloids with tranquilizing action. *Science* **123**:992–993.

Brodie, B. B., Olin, J. S., Kuntzman, R., and Shore, P. A. (1957a). Possible interrelationships between release of brain norepinephrine and serotonin by reserpine. *Science* **125**:1293–1294.

Brodie, B. B., Tomich, E. G., Kuntzman, R., and Shore, P. A. (1957b). On the mechanism of action of reserpine—effect of reserpine on capacity of tissues to bind serotonin. *J. Pharm. Pharmacol.* **119**:416–467.

Brodie, B. B., Spector, S., Kuntzman, R., and Shore, P. A. (1958). Rapid biosynthesis of brain serotonin before and after reserpine administration. *Naturwissenschaften* **45**:243.

Brodie, B. B., Spector, S., and Shore, P. A. (1959). Interaction of drugs with norepinephrine in the brain. *Pharmacol. Rev.* **11**:548–564.

Brodie, B. B., Maickel, R. P., and Westerman, E. O. (1961). Action of reserpine on pituitary–adrenocortical system through possible action on hypothalamus. In Kety, S. S., and Elkes, J. (eds.), *Regional Neurochemistry*, Pergamon Press, Oxford, pp. 351–361.

Brodie, B. B., Kuntzman, R., Hirsch, C. W., and Costa, E. (1962). Effects of decarboxylase inhibition on the biosynthesis of brain monoamines. *Life Sci.* **3**:81–84.

Brodie, B. B., Comer, M. S., Costa, E., and Dlabac, A. (1966a). The role of brain serotonin in the mechanism of the central action of reserpine. *J. Pharmacol. Exptl. Therap.* **152**: 340–349.

Brodie, B. B., Costa, E., and Dlabac, A. (1966b). Application of steady state kinetics to the estimation of synthesis rate and turnover time of tissue catecholamines. *J. Pharmacol. Exptl. Therap.* **154**:493–498.

Brooks, D. C. (1968). Localization and characteristics of the cortical waves associated with eye movement in the cat. *Exp. Neurol.* **22**:603.

Brown, G. M., and Hornykiewicz, O. (1971). Hypothalamic and median eminence catecholamines and thyroid functions. *The Society for Neuroscience First Annual Meeting,* Washington, D. C., October 27–30.

Brownlee, G., and Spriggs, T. L. B. (1965). Estimation of dopamine, noradrenaline, adrenaline and hydroxytryptamine from single rat brains. *J. Pharm. Pharmacol.* **17**:429–433.

Brune, G. G. (1965). Biogenic amines in mental illness. In Pfeiffer, C. C., and Smythies, J. R. (eds.), *International Review Neurobiology,* Academic Press, New York, pp. 197–220.

Brune, G. G., and Himwich, H. E. (1963). Biogenic amines and behavior in schizophrenic patients. *Recent Adv. Biol. Psychiat.* **5**:144–160.

Bruno, A., and Brigida, E. (1965). Azione della l-dopa sulla sintomatologia extrapiramidale da haloperidol. *Riv. Neurobiol.* **11**:646–654.

Bulat, M., and Supek, Z. (1967). The penetration of 5-hydroxytryptamine through the blood–brain barrier. *J. Neurochem.* **14**:265–271.

Bumpus, M., and Page, I. H. (1955). Serotonin and its methylated derivatives in human urine. *J. Biol. Chem.* **212**:111.

Bunney, W. E., Jr., Davis, J. M., and Weil-Malherbe, H. (1967). Biochemical changes in psychotic depression. High norepinephrine levels in psychotic vs. neurotic depression. *Arch. Gen. Psychiat.* **16**:448–460.

Burack, W. R., and Draskoczy, P. R. (1964). The turnover of endogenously labeled catecholamines in several regions of the sympathetic nervous system. *J. Pharmacol. Exptl. Therap.* **144**:66–75.

Burger, M. (1957). Veränderungen der Adrenalin und Noradrenalin Konzentration im menschlichen Blut Plasma unter Reserpin. *Arch. Exptl. Pathol. Pharmacol.* **230**:489–498.

Burgermeister, J. J., Dick, P., Garonne, G., Guggisberg, M., and Tissot, R. (1963). Urinary excretion of 5-hydroxyindolacetic acid (5-HIAA) in 150 patients with depressive syndrome and maniacal agitation (its modifications by 5-hydroxytryptophan loading and therapy in the depressive states). *Presse Méd.* **71**:1116–1118.

Burkard, W.P. (1962). A new inhibitor of decarboxylase of aromatic amino acids, *Experientia* **18**:411.

Burkard, W. P., Gey, K. F., and Pletscher, A. (1962). A new inhibitor of decarboxylase of aromatic amino acids. *Experientia* **18**:1–5.

Burn, J. H. (1966). Adrenergic transmission. *Pharmacol. Rev.* **18**:459–470.

Buscaino, G. A., and Stefanachi, L. (1958). Contributo al metabolismo della sostanze indolich con particolare reguardo all 5-HIAA nella schizofrenia. *Confinia Neurol. (Basel)* **18**: 188–195.

Cahn, J., and Herold, M. (1960). Etude pharmacologique du Ro 1-9569 (tetrabenazine). *Psychiat. Neurol. (Basel)* **140**:210–215.

Cahn, J., and Herold, M. (1964). Le rôle des catécholamines dans le contrôle de la neurosecretion hypothalamique. *Agressologie* **5**:451–463.

Cahn, J., Herold, M., Dubrasquet, M., Barre, N., Alano, J., and Breton, Y. (1958). Contribution à un concept biochimique des psychoses experimentales VIII. Action de l'iproniazid pervitine, LSD sur le bilan humoral et le métabolisme cérébral du lapin vigil. *Compt. Rend. Soc. Biol. (Paris)* **152**:1479–1481.

Callingham, B. A., and Cass, R. (1963). A modification of the butanol excretion method for the fluorimetric assay of catecholamines in biological materials. *J. Pharm. Pharmacol.* **15**: 699–700.

Campus, S., Accation, G., and Rappelli, A. (1965). Escrezione urinaria di catecolamine e variazioni tensive in soggeti normotesi durante calcolo mentales. *Boll. Soc. Ital. Biol. Sper.* **41**:9–11.

Canal, N., and Ornesi, A. (1961). Serotonina encefalica e ipertermia de vaccine. *Atti. Accad. Med. Lombarda* **16**:65–69.

Cannon, W. B. (1915). *Bodily Changes in Pain, Hunger, Fear and Rage*, Appelton, New York.

Carlsson, A. (1959). The occurrence, distribution and physiological role of catecholamines in the nervous system. *Pharmacol. Rev.* **11**:490–493.

Carlsson, A. (1961). Brain monoamines and psychotropic drugs. In Rothlin, E. (ed.), *Neuropsychopharmacology*, Elsevier, Amsterdam, pp. 417–421.

Carlsson, A. (1964). Functional significance of drug-induced changes in monoamine levels. *Progr. Brain Res.* **8**:9–27.

Carlsson, A. (1965). Drugs which block the storage of 5-hydroxytryptamine and related amines. In Erspamer, V. (ed.), *Handbuch Exptl. Pharmacol.*, Springer-Verlag, Berlin, pp. 529–592.

Carlsson, A. (1966). Modification of sympathetic function. Pharmacological depletion of catecholamine stores. *Pharmacol. Rev.* **18**:541–549.

Carlsson, A., and Corrodi, H. (1965). In den Catecholamine-Metabolismus eingreifende Substanzen, 3) 2,3-Dihydroxyphenylacetamide und verwandte Verbindungen. *Helv. Chim. Acta* **47**:1340–1349.

Carlsson, A., and Hillarp, N. A. (1962). Formation of phenolic acids in brain after administration of 3,4-dihydroxyohenylalanine. *Acta Physiol. Scand.* **55**:95–105.

Carlsson, A., and Lindqvist, M. (1961). Dopa analogues as tools for the study of dopamine and noradrenaline in brain. In Ajuriaguerra, J. de (ed.), *Monoamines et système nerveux central*, Symposium Bel-Air, Georg & Cie, Geneva.

Carlsson, A., and Lindqvist, M. (1962). A method for the determination of normetanephrine in brain. *Acta Physiol. Scand.* **54**:83–86.

Carlsson, A., and Lindqvist, M. (1963). Effect of chlorpromazine or haloperidol on formation of 3-methoxytyramine and normetanephrine in mouse brain. *Acta Pharmacol. Toxical.* **20**:140–144.

Carlsson, A., and Waldeck, B. (1965). Mechanism of amine transport in the cell membranes of the adrenergic nerves. *Acta Pharmacol.* **22**:293–300.

Carlsson, A., Lindqvist, M., and Magnusson, T. (1957*a*). 3,4-Dihydroxyphenylalanine and 5-hydroxytryptophan as reserpine antagonists. *Nature* **180**:1200.

Carlsson, A., Rosengren, E., Bertler, A., and Nilsson, J. (1957*b*). Effect of reserpine on the metabolism of catecholamines. In Garattini, S., and Ghetti, V. (eds.), *Psychotropic Drugs*, Elsevier, Amsterdam, pp. 363–372.

Carlsson, A., Lindqvist, M., Magnusson, T., and Waldeck, B. (1958). On the presence of 3-hydroxytryptamine in brain. *Science* **127**:471.

Carlsson, A., Lindqvist, M., and Magnusson, T. (1959). The effect of MAO inhibitors on the metabolism of the brain catecholamine. *J. Soc. Cienc. Med. Lisboa*, Suppl. 123, pp. 96–98.

Carlsson, A., Lindquist, M., and Magnusson, T. (1960). On the biochemistry and possible functions of dopamine and noradrenaline in brain. In Vane, R. J., Wolestenheim, G. W., and O'Connor, M. (eds.), *Adrenergic Mechanisms* (Ciba Foundation Symposium), J. A. Churchill Ltd., London, pp. 432–439.

Carlsson, A., Falck, B., Hillarp, N. A., and Thieme, G. (1961). A new histochemical method for visualization of tissue catecholamines. *Med. Exp.* **4**:123–124.

Carlsson, A., Falck, B., and Hillarp, N. A. (1962*a*). Cellular localization of brain monoamines. *Acta Physiol. Scand.* **56**: Suppl. 196.

Carlsson, A., Falck, B., Hillarp, N. A. and Torp, A. (1962*b*). Histochemical localization at the cellular level of hypothalamic noradrenaline. *Acta Physiol. Scand.* **54**:385–386.

Carlsson, A., Hillarp, N. A., and Waldeck, B. (1963). Analysis of Mg^{++}-ATP dependent storage mechanism in the amine granules of the adrenal medulla. *Acta Physiol. Scand.* **59**: (Suppl. 215) 1–38.

Carlsson, A., Falck, B., Fuxe, K., and Hillarp, N. A. (1964). Cellular localization of monoamines in the spinal cord. *Acta Physiol. Scand.* **60**:112–119.

Carlsson, A., Dahlström, A., Fuxe, K., and Hillarp, N. A. (1965*a*). Failure of reserpine to deplete noradrenaline neurons of alpha-methyl-noradrenaline formed from alpha-methyl-DOPA. *Acta Pharmacol.* **22**:270–276.

Carlsson, A., Dahlström, A., Fuxe, K., and Lindqvist, M. (1965*b*). Histochemical and bio-chemical detection of monoamine release from brain neurons. *Life Sci.* **4**:809–816.

Carlyle, A., and Grayson, J. (1956). Factors involved in the control of cerebral blood flow. *J. Physiol. (Lond.)* **133**:10.

Cazullo, C. L. (1960). Azione degli inhibitori delle monoaminossidasi e della imipramina nella depressione: Psicodinamica et reattivita biologica. Symp. le Sindromi Depressive, Min. Med. Ed, Torino, Atti 123.

Cazzullo, C. L. (1967). Pharmacologic and clinical differences in the action of sedative-hypnotic tranquilizers and neuroleptic drugs. In Brill, H., (ed.), *Neuropsychopharmacology,* Excerpta Medica Foundation, Amsterdam, pp. 124–129.

Cerletti, A., and Rothlin, E. (1955). Role of 5-HT in mental diseases and its antagonism to lysergic acid derivatives. *Nature* **176**:785–786.

Cerletti, A., Taeschler, M., and Weidmann, H. (1968). Pharmacologic studies on the structure-activity relationship of hydroxyindole alkylamines. *Adv. Pharmacol.* **6B**:233–246.

Chase, T. H., Breese, T. R., and Kopin, I. J. (1967). Serotonin release from brain by electrical stimulation: Regional differences and effect of LSD. *Science* **157**:1461–1463.

Chen, G., Ensor, C. R., and Bohner, B. (1968). Drug effects on the disposition of active biogenic amines in the CNS. *Life Sci.* **7**:1063–1074.

Chessick, R. D. (1966). A comparison of the effect of infused catecholamines and certain affect states *Am. J. Psychiat.* **123**:156–165.

Chiesa, F. (1959). Brain serotonin increase in rats and cats after treatment with neuroketone (a brain extract). *Enzymologia* **21**:61–66.

Clouet, D. H., and Ratner, M. (1970). Catecholamine biosynthesis in brains of rats treated with morphine. *Science* **168**:854.

Cohen, R. A., Bridgers, R. F., Axelrod, J., Weil-Malherbe, H., La Brosse, E. H., Bunney, W. E., *et al* (1968). The metabolism of the catecholamines (clinical implications). *Ann. Int. Med.* **56**:960–987.

Cohn, C. K., Dunner, D. L., and Axelrod, J. (1970). Reduced COMT activity in red blood cells of women with primary affective disorders. *Science* **170**:1323–1324.

Constantinidis, J. (1965). L'excretion urinaite des acides aminés dans les oligophrenies endogènes. *Arch. Suisses Neurol. Neurochirugie. Psychiat.* **96**:1–49.

Constantinidis, J., Bartholini, G., Tissot, R., and Pletscher, A. (1967). Elektive Anreicherung von Dopamin im Parenchym des Ratterhirns. *Helv. Physiol. Pharmacol. Acta* **25**:411.

Constantinidis, J., Bartholini, G., Tissot, R., and Pletscher, A. (1968). Accumulation of dopa-mine in the parenchyma after decarboxylase inhibition in the capillaries of brain. *Experientia* **24**: Separatum 13.

Constantinidis, J., de la Torre, J. C., Tissot, R., and Geissbuhler, F. (1969*a*). La barrière cap-illaire pour la dopa dans le cerveau et les differents organes. *Psychopharmacologia* **15**:75–87.

Constantinidis, J., Tissot, R., and de la Torre, J. C. (1969*b*), Essai de localisation des mono-amines cérébrales dans l'hypothalamus humain. *Pathol. Biol.* **17**:361–362.

Constantinidis, J., de la Torre, J. C., Tissot, R., and Huggel, H. (1970). Les monoamines cérébrales lors de l'hibernation chez la chauve-souris. *Rev. Suis. Zool.* **77**:345–352.

Constantinidis, J., Geissbuhler, F., and Tissot, R. (1971). Histochemical study on the effect of intra-ventricular administration of 6-hydroxydopamine on monoamine-containing neurons in the CNS. In Malmfors, T. (ed.), *6-Hydroxydopamine and CNS*, in press.

Cooper, K. E. (1966). Temperature regulation in the hypothalamus. *Brit. Med. Bull.* **22**:238–242.

Cooper, J. R., and Mercer, I. (1961). The enzymic oxidation of tryptophan to 5-hydroxy-tryptophan in the biosynthesis of serotonin. *J. Pharm. Exp. Ther.* **132**:265–268.

Cooper, K. E., Cranston, W. I., and Honour, A. J. (1964). Temperature changes induced by

5-HT noradrenaline and pyrogens injected into the rabbit brain. *J. Physiol. (Lond.)* **175**:68.

Correale, P. (1956). The occurrence and distribution of 5-hydroxytryptamine (enteramine) in the central nervous system of vertebrates. *J. Neurochem.* **1**:22–31.

Corrodi, H., Falck, B., and Hillarp, N. A. (1962). Sensitive flourescence methods for histochemical demonstration of catecholamine at the cellular level. Report at the meeting for Scandanavian Pharmacologists, August, Göteborg, Sweden.

Corrodi, H., Fuxe, K., and Schou, M. (1969). The effect of prolonged lithium administration on cerebral monoamine neurons in the rat. *Life Sci.* **8**:643–651.

Costa, E., and Aprison, M. H. (1957). Fate of serotonin perfused through an isolated brain with intact nervous connections. *Fed. Proc.* **16**:25.

Costa, E., and Aprison, M. H. (1958). Studies on the 5-hydroxytryptamine (serotonin) content in human brain. *J. Nerv. Ment. Dis.* **126**:289–293.

Costa, E., Revzin, A. M., Kuntzman, R., Spector, S., and Brodie, B. B. (1961). Role for ganglionic norepinephrine in sympathetic synaptic transmission. *Science* **133**:1822–1823.

Cotzias, G. C., Van Woert, M. H., and Schiffer, L. M. (1967). Aromatic amino acids and modification of Parkinsonism. *New Engl. J. Med.* **276**:374.

Coury, J. N. (1967). Neural correlates of food and water intake in the rat. *Science* **156**:1763–1765.

Cox, R. H., and Maickel, R. P. (1969). Effects of guanethidine on rat brain serotonin and norepinephrine. *Life Sci.* **8**:1319–1324.

Craigie, E. H. (1920). On the relative vascularity of various parts of the central nervous system of the albino rat. *J. Comp. Neurol.* **31**:429–464.

Craigie, E. H. (1940). Measurement of vascularity in some hypothalamic nuclei of the albino rat. *Res. Publ. Ass. Nerv. Ment. Dis.* **20**:310–319.

Crawford, T. B. B. (1957). The distribution of 5-HT in the central nervous system of the dog. In *5-HT,* Pergamon Press, Oxford, pp. 20–25.

Cremata, V. Y., and Koe, B. K. (1966). Clinical–pharmacological evaluation of *p*-chlorophenylalanine: A new serotonin-depleting agent. *Clin. Pharmacol. Therap.* **7**:768–776.

Creveling, C. R., Levitt, M., and Udenfriend, S. (1962). An alternative route for biosynthesis of norepinephrine. *Life Sci.* **10**:523.

Cross, P. A., and Silver, I. A. (1966). Electrophysiological studies on the hypothalamus. *Brit. Med. Bull.* **22**:254–260.

Crout, R. J. (1961). Effect of inhibiting both catechol-*O*-methyltransferase and monoamine oxidase on cardiovascular responses to norepinephrine. *Proc. Soc. Exptl. Biol.* **108**:482–484.

Culley, J., Saunder, R. N., Mertz, E. T., and Jolly, D. H. (1963). Effect of a tryptophan deficient diet on brain serotonin and plasma tryptophan level. *Proc. Soc. Explt. Biol.* **113**:645–648.

Curtis, D. R. (1964). Microelectrophoresis. In Nastuk, W. (ed.), *Physical Techniques in Biological Research,* Academic Press, New York, Vol. 5, pp. 144–190.

Curzon, G., and Green, A. R. (1968). Effect of hydrocortisone on rat brain 5-hydroxytryptamine. *Life Sci.* **7**:657–663.

Dagirmanjian, R. (1963). The effects of guanethidine on the noradrenaline content of the hypothalamus in the cat and rat. *J. Pharm. Pharmacol.* **15**:518–521.

Dahl, E., Falck, B., Lindqvist, M., and Mecklenburg, C. (1962). Monoamines in mollusk neurons. *Kgl. Fisiogr. Sallsk. Hdl.* **32**:89–92.

Dahlström, A., and Fuxe, K. (1964*a*). Evidence for the existence of monoamine containing neurons in the central nervous system. I. Demonstration of monoamines in the cell bodies of brain stem neurons. *Acta. Physiol. Scand.* **62**:(Suppl. 232) 1–55.

Dahlström, A., and Fuxe, K. (1964*b*). Localization of monoamines in the lower brain stem. *Experientia* **20**:398–399.

Dahlström, A., and Fuxe, K. (1964*c*). A method for the demonstration of monoamine containing nerve fibers in the central nervous system. *Acta Physiol. Scand.* **60**:293–294.

Dahlström, A., and Fuxe, K. (1965a). Evidence for the existence of an outflow of noradrenaline nerve fibres in the ventral roots of the rat spinal cord. *Experientia* **21**:409–410.

Dahlström, A., and Fuxe, K. (1965b). Evidence for the existence of monoamine neurons in the central nervous system. II. Experimentally induced changes in the intra-neuronal amine levels of bulbospinal neuron system. *Acta Physiol. Scand.* **64**: Suppl. 247.

Dahlström, A., Fuxe, K., Olson, L., and Ungerstedt, U. (1964). Ascending systems of catecholamine neurons from lower brain stem. *Acta Physiol. Scand.* **62**:485–486.

Dahlström, A., Fuxe, K., and Hillarp, N. A. (1965). Site action of reserpine. *Acta Pharmacol. (Kbh.)* **22**:277–292.

Dahlström, A., Häggendahl, J., and Hökfelt, T. (1966). The noradrenaline content of the varicosities of sympathetic adrenergic nerve terminals in the rat. *Acta Physiol. Scand.* **67**:271–277.

Da Prada, M., and Pletscher, A. (1966). Acceleration of the cerebral dopamine turnover by chlorpromazine. *Experientia* **22**:465–466.

Dasgupta, S. R., and Werner, G. (1954). Inhibition of hypothalamic, medullary and reflex vasomotor responses by chlorpromazine. *Brit. J. Pharmacol.* **9**:389–391.

Davis, V., and Walsh, M. J. (1970). Alcohol, amines and alkaloids: A possible biochemical basis for alcohol addiction. *Science* **167**:1005–1006.

Debijadji, R., Varagic, V., Dekleva, N., Elcic, S., Stefanovic, M., Davidovic, J., and Marisavijevic. T. (1965). The effect of severe hypoxic hypoxia in the decompression chamber on the catecholamine content of the hypothalamus in the cat. *Experientia* **21**:153.

De Caro, G. (1964). 5-Hydroxytryptamine, metanephrine and histamine in the brain of *Rana esculenta. J. Neurochem.* **11**:825.

De la Lande, I. S., and Harvey, J. A. (1965). A new and sensitive bioassay for catecholamines. *J. Pharm. Pharmacol.* **17**:589–593.

de la Torre, J. C. (1968a). Effect of LSD-25 on the septal region of the rat brain. *Nature* **219**:954–955.

de la Torre, J. C. (1968b). Monoamines cérébrales avec référence particulière à l'hypothalamus. Méd. Hyg. Monograph No. 1477, pp. 1–16.

de la Torre, J. C. (1970a). Penetration of monoamine precursors through the blood–brain barrier using DMSO. *Ann. Int. Med.* **75**:806 (abs).

de la Torre, J. C. (1970b). Relative penetration of L-dopa and 5-HTP through the brain barrier using dimethyl sulfoxide. *Experientia* **26**:1117–1118.

de la Torre, J. C. (1971). The blood–brain barrier for L-dopa in the hypothalamus. *J. Neurol. Sci.* **12**:77–93.

de la Torre, J. C., and Boggan, W. O. (1971). A common error in the measurement of brain dopamine following L-dopa. The Society for Neuroscience First Annual Meeting, Washington, D. C., October 27–30, 1971.

de la Torre, J. C., and Mullan, S. (1970). A possible role for serotonin in drug-induced seizures. *J. Pharm. Pharmacol.* **22**:858–859.

de la Torre, J. C., and Mullan, S. (1971). Blood–brain barrier mechanisms of L-dopa. *Neurology* **21**:446.

de la Torre, J. C., and Mullan, S. (1972). Experimental evaluation of L-dopa penetration in brain. *Trans. Am. Neurol. Assoc.,* **96**:227–229.

de la Torre, J. C., Kawanaga, H., and Mullan, S. (1970). Seizure susceptibility after manipulation of brain serotonin. *Arch. Int. Pharmacodyn.* **188**:298–304.

Delgado, J. M. R. (1960). Emotional behavior in animals and humans. *Psychiat. Res.* **12**:259.

Delorme, F. (1966). *Monoamines et sommeils. Etude polygraphique, neuropharmacologique et histochimique des états de sommeil chez le chat,* Imprimerie L. M. C., Lyon.

Delorme, F., Froment, J. L., and Jouvet, M. (1966). Suppression du sommeil par la *p*-chlorométhamphétamine et la *p*-chlorophénylalanine. *C. R. Soc. Biol.* **160**:2347.

DeMaio, D. (1963). Influence of adrenalactomy and hypophysectomy on cerebral serotonin. *Science* **129**:1678–1679.

DeMaio, D., and Paquariello, G. (1963). Gamma-amino-beta-hydroxybutyric acid (GABOB) and brain serotonin. *Psychopharmacologia* **5**:84–86.

Dement, W. (1966). Psychiatric implications of dream deprivation. Paper presented at a at a symposium, *Psychodynamic Implications of Physiological Studies in Dreams*. Women's Medical College, Philadelphia.

Dement, W., and Kleitman, N. (1957). Cyclic variations in EEG during sleep and their relation to eye movements, body motility and dreaming. *Electroenceph. Clin. Neurophysiol.* **9**:673–690.

Demetrescu, M. (1965). Some relationships between the hypothalamus and the reticular formation. *Stud. Cercet. Fiziol.* **10**:49–53.

Demis, D. J., Blaschko, H., and Welch, A. D. (1955). The conversion of dihydroxyphenyl-alanine-2-C^{14} (dopa) to norephrinephrine by bovine adrenal medullary homogenates. *J. Pharmacol. Exptl. Therap.* **113**:114.

Dengler, H. J., Spiegel, H. E., and Titus, E. C. (1961). Uptake of tritium labeled norepinephrine in brain and other tissues of the cat *in vitro. Science* **133**:1072–1073.

Dengler, H. J., Michaelson, I. A., Spiegel, H. E., and Titus, E. (1962). The uptake of labeled norepinephrine by isolated brain and other tissues of the cat. *Internat. J. Neuropharmacol.* **1**:23.

Denko, C. W., Goodman, R. M., Miller, R., and Donovan, T. (1967). Distribution of dimethyl sulfoxide- S in the rat. *Ann. N. Y. Acad. Sci.* **141**:77–84.

Denny-Brown, D. (1954). In Doshay, L. J. (ed.), *Parkinsonism and Its Treatment*, Lippincott, Philadelphia.

De Robertis, E. (1965). Synaptic vesicles from the rat hypothalamus. Isolation norepinephrine content. *Life Sci.* **4**:193–201.

De Robertis, E. (1966). Adrenergic endings and vesicles isolated from brain. *Pharmacol. Rev.* **18**:413–434.

De Robertis, E. (1967). Ultrastructure and cytochemistry of the synaptic region. *Science* **156**:907–914.

De Robertis, E., and Bennett, H. S. (1955). Some features of the submicroscopic morphology of synapses in frog and earth-worm. *J. Biophys. Biochem. Cytol.* **1**:47–58.

De Robertis, E., and Pellegrino de Iraldi, A. (1961). Plurivesicular secretory processes and nerve endings in pineal gland of rat. *J. Biophys. Biochem. Cytol.* **10**:361–372.

De Robertis, E., Pellegrino de Iraldi, A., Rodriguez de Lores Arnaiz, G., and Gomez, G. J. (1960). Aislamiento de terminaciones nerviosas y vesiculas sinápticas. *Sesiones Soc. Arg. Biol., Mendoza*, pp. 24–25.

De Robertis, E., Rodriguez de Lores Arnaiz, G., and Pellegrino de Iraldi, A. (1962). Isolation of synaptic vesicles from nerve endings of the rat brain. *Nature* **194**:794–795.

De Robertis, E., Rodriguez de Lores Arnaiz, G., and Alberici, M. (1969). Ultra-structural neurochemistry. In Jasper, H. H., Ward, A. A., and Pope, A. P. (eds.), *Basic Mechanisms of the Epilepsies*. Little, Brown, and Co., Boston, 137–158.

Detoni, G. (1953). I movimenti pendolari dei bulbi oculari dei bambini durante il sommo fisiologico ed in alcuni stati morbosi. *Pediatria* **41**:489.

Dewhurst, W. G. (1968). Methysergide in mania. *Nature* **219**:506–507.

Donaldson, R. M., Arabethy, J., and Gray, S. J. (1959), 5-hydroxyindole metabolism in patients with hepatic cirrhosis and in rats. *J. Clin. Invest.* **38**:933.

Donoso, A. O. (1964). Cortical "spreading depression" and hypothalamic noradrenaline. *Acta Physiol. Lat-Am.* **14**:399–400.

Donoso, A. O. (1966). Effect of brain lesions on hypothalamic noradrenaline in rats. *Experientia* **22**:191.

Donoso, A. O., Stefano, F. J., and Biscardi, A. M. (1967). Effect of castration on hypothalamic catecholamines. *Am. J. Physiol.* **212**:737–739.

Donovan, B. T. (1966). Experimental lesions of the hypothalamus. A critical survey with particular reference to endocrine effects. *Brit. Med. Bull.* **22**:249–253.

Draskoczy, P. R., and Lyman, C. P. (1967). Turnover of catecholamines in active hibernating ground squirrels. *J. Pharmacol. Exptl. Therap.* **155**:101–111.

Dubnick, B., Leesen, H. A., Chessin, M., and Scolt, C. C. (1960). Accumulation of serotonin in the brain of reserpined heated mice after inhibition of MAO. Effect of body temperature. *Arch. Int. Pharmacodyn.* **126**:194–202.

Dubnick, B., Leeson, G. A., and Phillips, G. E. (1962). *In vivo*-inhibition of serotonin synthesis in mouse brain by β-phenylethylhydrazine, an inhibitor of monoamine oxidase. *Biochem. Pharmacol.* **11**:45–52.

Duchesne, P. Y. (1964). Mise en évidence histochimique de catécholamines ou d'indolamines dans la glande pinéale du rat et tentative d'indentification. *Compt. Rend. Soc. Biol. (Paris)* **158**:1174–1176.

Duchesne, P. Y. (1966). Attempt at localization and histochemical identification of cerebral biogenic amines. *Acta Neurol. Belg.* **66**:467–478.

Duvoisin, R., Yahr, M. D., Schear, M., Hoehn, M. M., and Barrett, R. E. (1969). The present status of L-dopa in the treatment of parkinsonism. Ninth International Congress of Neurology, New York, September 22–27, 1969.

Eccles, J. C. (1957). *The Physiology of the Nerve Cell,* The Johns Hopkins Press, Baltimore.

Eccles, J. C. (1964). *The Physiology of Synapses,* Springer, Berlin.

Edelman, P. M., Schwartz, I. L., Cronkite, E. P., and Livingston, L. (1965). Studies of the ventro-medial hypothalamus with autoradiographic techniques. *Ann. N. Y. Acad. Sci.* **131**: 485–501.

Edström, R. (1964). The blood–brain barrier concept. *Internat. Rev. Neurobiol.* **7**:153–190.

Ehringer, H., and Hornykiewicz, O. (1960). Verteilung von Noradrenalin und Dopamin (3-Hydroxytyramin) im Gehirn des Menschen und ihr Verhalten bei Erkrankungen des extrapyramidalen Systems. *Klin. Wschr.* **38**:1236–1239.

Eidelberg, E., Miller, M. K., and Long, M. (1966). Spectrum analysis of electroencephalographic changes induced by some psychoactive agents. Their possible relationship to changes in cerebral biogenic amine levels. *Internat. J. Neuropharmacol.* **5**:59–74.

Eisenfeld, A. J., Krakoff, L., Iversen, L. L., and Axelrod, J. (1967). Inhibition of the extraneuronal metabolism of noradrenaline in the isolated heart by adrenergic blocking agents. *Nature (Lond.)* **213**:297–298.

Elvidge, A. R., and Reed, G. E. (1938). Biopsy studies of cerebral pathologic changes in schizophrenia and manic-depressive psychosis. *Arch. Neurol. Psychiat.* **40**:227.

Elmadjian, F., Hope, J. M., and Lamson, E. T. (1957). Excretion of epinephrine in various emotional states. *J. Clin. Endocrinol.* **17**:608–620.

Eränkö, O. (1955). Distribution of fluorescing islets, adrenaline and noradrenaline in the adrenal medulla of the cat. *Acta Endocrinol. (Kbh.)* **18**:180–188.

Eränkö, O. (1964). Histochemical demonstration of catecholamines by fluorescence induced by formaldehyde vapour. *J. Histochem. Cytochem.* **12**:487–489.

Eränkö, O. (1966). Demonstration of catecholamines and cholinesterases in the same section. *Pharmacol. Rev.* **18**:353–358.

Eränkö, O. (1967). Histochemistry of nervous tissues: Catecholamines and cholinesterases. *Ann. Rev. Pharmacol.* **7**:203–222.

Ernsting, M. J., Kafoe, W. F., Nauta, W. T., Oosterhuis, H. K., and Roukema, P. A. (1962). Investigation into the effect of orphenadrine hydrochloride (Disipal) on the monoamine oxidase in the brain and liver of rats and guinea pigs. *Med. Exp.* **7**:119–124.

Erspamer, V. (1955). Observations on the fate of indolalkylamines in the organism. *J. Physiol.* **127**:118.

Eskin, I. A. (1964). Peculiarities of catecholamine metabolism in the central nervous system in states of stress. *Dokl. Akad. Nauk SSSR* **1559**:693–695.

Euler, U. S. von (1946). A specific sympathomimetic ergone in adrenergic nerve fibres (sympathin) and its relations to adrenaline and noradrenaline. *Acta Physiol. Scand.* **12**: 73–97.

Euler, U. S. von (1951). Excretion of noradrenaline, adrenaline and hydroxytyramine. *Acta Physiol. Scand.* **22**:1.

Euler, U. S. von (1961). Neurotransmission in the adrenergic nervous system. *Harvey Lect.* *(N. Y.)* **55**:65.

Euler, U. S. von (1964). Aufnahme, Speicherung und Feisetzung von Katecholaminen in adrenergischen Neuronen. *Pflügers Arch. Ges. Physiol.* **281**:1–3.

Euler, U. S. von, and Lishajko, F. (1959). Excretion of catechol in human urine. *Nature* **183**: 1123.

Euler, U. S. von, and Lishajko, F. (1961). Uptake of catecholamines by adrenergic nerve granules. *Acta Physiol. Scand.* **53**:196.

Euler, U. S. von, and Lundberg, U. N. (1954). Effect of flying on the epinephrine excretion in air force personnel. *J. Appl. Physiol.* **6**:551.

Euler, U. S. von, Lishajko, F., and Stjarne, L. (1963). Catecholamines and adenosine triphosphate in isolated adrenergic nerve granules. *Acta Physiol. Scand.* **59**:495–496.

Evarts, V. E. (1957). A review of the neurophysiological effects of lysergic acid diethylamide (LSD) and other psychotomimetic agents. *Ann N. Y. Acad. Sci.* **66**:479–495.

Everett, J. W., and Quinn, D. L. (1966). Differential hypothalamic mechanisms inciting ovulation and pseudopregnancy in the rat. *Endocrinology* **78**:141–150.

Falck, B. (1962). Observations on the possibilities for the cellular localization of monoamines with a fluorescence method. *Acta Physiol. Scand.,* Suppl. 197.

Falck, B. (1964). Cellular localization of monoamines. In: Himwich, H., and Himwich, W. (eds.), *Biogenic amines. Progr. Brain Res.* **8**:26–44.

Falck, B., and Hillarp, N. D. (1959). On the cellular localization of catechol in the brain. *Acta Anat.* **38**:277–279.

Falck, B., and Owman, C. (1965). A detailed methodological description of the fluorescence method for the cellular demonstration of biogenic monoamines. *Acta Univ. Lund* **2**: No. 7.

Falck, B., and Torp, A. (1962a). A fluorescence method for histochemical demonstration of noradrenaline in the adrenal medulla. *Med. Exp.* **5**:429–432.

Falck, B., and Torp, A. (1962b). A new evidence for the localization of noradrenaline in adrenergic nerve terminals. *Med. Exp.* **6**:1269–1272.

Falck, B., Hillarp, N. A., Thieme, G., and Torp, A. (1962). Fluorescence of catecholamines and related compounds condensed with formaldehyde. *J. Histochem. Cytochem.* **10**: 348–354.

Faurbye, A., and Pind, K. (1964). Investigation on the occurrence of the dopamine metabolite 3,4-dimethoxyphenylethylamine in the urine of schizophrenics. *Acta Psychiat. Scand.* **40**:240–243.

Feldberg, W. (1958). Anaesthesia and sleep-like conditions produced by injections into cerebral ventricles of the cat. *J. Physiol. (Lond.)* **140**:20.

Feldberg, W. (1965). A new concept of temperature control in the hypothalamus. *Proc. Roy. Soc. Med.* **58**:395–404.

Feldberg, W., and Myers, R. D. (1963). A new concept of temperature regulation by amines in the hypothalamus. *Nature* **200**:1325 ff.

Feldberg, W., and Myers, R. D. (1964). Temperature changes produced by amines injected into the cerebral ventricles during anaesthesia. *J. Physiol. (Lond.)* **175**:464–478.

Feldberg, W., and Myers, R. D. (1965a). Changes in temperature produced by micro-injections of amines into the anterior hypothalamus of cats. *J. Physiol. (Lond.)* **177**:239–245.

Feldberg, W., and Myers, R. D. (1965b). Hypothermia induced by chlorase acting on hypothalamus. *J. Physiol. (Lond.)* **179**:509–517.

Feldberg, W., Hellon, R. F., and Myers, R. D. (1966). Effects on temperature of monoamines injected into the cerebral ventricles of anaesthetized dogs. *J. Physiol. (Lond.)* **186**:416.

Feldstein, A., Hoagland, H., and Freeman, H. (1958). On the relation of serotonin to schizophrenia. *Science* **128**:358–359.

Fieve, R. R., Platman, S. R., and Fleiss, J. L. (1969). A clinical trial of methysergide and lithium in mania. *Psychopharmacologia* **15**:425–429.

Fischer, J. E., Kopin, I. J., and Axelrod, J. (1965). Evidence for extra-neuronal binding of norepinephrine. *J. Pharmacol. Exptl. Therap.* **147**:181–185.

Fleisschhacker, H. H., Lancaster, J. B., and Wheeler, A. B. (1959). Studies in schizophrenia. Disturbance of protein and tryptophan metabolism and toxicity of urine. *J. Ment. Sci.* **105**:313–325.

Fleming, R. M. (1965). Single extraction method for the simultaneous fluorometric determination of serotonin, dopamine, and norepinephrine in brain. *Anal. Chem.* **37**:692–696.

Florey, E., and McLennan, H. (1955). The release of an inhibitory substance from mammalian brain, and its effect on peripheral synaptic transmission. *J. Physiol. (Lond.)* **129**: 384–392.

Folklow, B., and Rubinstein, E. H. (1966). The functional role of some autonomic and behavioral patterns evoked from the lateral hypothalamus of the cat. *Acta Physiol. Scand.* **66**:182–188.

Fonberg, E. (1968). The role of the amygdaloid nucleus in animal behavior. In Astratyan, E. A. (ed.), *Brain Reflexes,* Elsevier, Amsterdam, pp. 273–281.

Frankenhaeuser, M. (1965). Interindividual differences in catecholamine excretion during stress. *Scand. J. Psychol.* **6**:117–123.

Frankenhaeuser, M., and Post, B. (1962). Catecholamine excretion during mental work as modified by centrally acting drugs. *Acta Physiol. Scand.* **55**:74–81.

Franzen, F., and Eysell, K. (1969). *Biologically Active Amines Found in Man,* Pergamon, Oxford, pp. 1–244.

Frazier, C. H., Alpers, B. J., and Lewy, F. H. (1936). The anatomical localization of the hypothalamic centre for the regulation of temperature. *Arch. Neurol. Psychiat.* **59**:122.

Freedman, D. X. (1963). Psychotomimetic drugs and brain biogenic amines. *Am. J. Psychiat.* **119**:843–850.

Freedman, D. X., and Giarman, N. J. (1962). LSD-25 and the status and level of brain serotonin. *Ann. N. Y. Acad. Sci.* **96**:98–106.

Fresia, P., Genovese, E., Valsecchi, A., and Valzelli, L. (1957). Variazioni nel contenuto della serotonina encefalica in differenti specie animali dopo elettroshock. *Boll. Soc. Ital. Biol. Sper.* **33**:888–890.

Freud, S. (1955). *The Interpretation of Dreams.* New York, Basic Books.

Frey, R. (1965). Research on postoperative and posttraumatic catecholamine levels. *Anesth. Analg. Curr. Res.* **22**:471–474.

Friedhoff, A. J., and Hollister, L. E. (1966). The role of catecholamines in specific psychiatric disorders. *Res. Publ. Ass. Res. Nerv. Ment. Dis.* **43**:366–370.

Friedhoff, A. J., and Van Winkle, E. (1963). Conversion of dopamine to 3,4-dimethoxyphenylacetic acid in schizophrenic patients. *Nature* **199**:1271–1272.

Fugassa, J. (1963). Contribution a l'étude de la noradrenaline cérébrale. *J. Physiol. (Paris)* **55**: (Suppl. 8) 1–76.

Fuller, R. W., Hines, C. W., and Mills, J. (1965). Lowering of brain serotonin level by chloramphetamines. *Biochem. Pharmacol.* **14**:483–488.

Fulton, J. F. (1951). *Frontal Lobotomy and Affective Behavior: A Neurophysiological Analysis,* W. W. Norton Company, New York.

Funatogawa, S. (1964). Methamphetamine-induced changes in behavior of cats in topographical distribution of brain serotonin. *Psychiat. Neurol. Jap.* **66**:743–754.

Funkenstein, D. H., King, S. H., and Drolette, M. (1954). The direction of anger during a laboratory stress-inducing situation. *Psychosom. Med.* **16**:404.

Furchgott, R. F., Kirpekar, S., Riker, M., and Schwab, A. (1963). Actions and interactions of norepinephrine, tyramine, and cocaine on aortic strips of rabbit and left atria of guinea pig and cat. *J. Pharmacol. Exptl. Therap.* **142**:39–58.

Fuxe, K. (1963). Cellular localization of monoamines in the median eminence and the infundibular stem of some mammals. *Acta Physiol. Scand.* **58**:383–384.

Fuxe, K. (1965a). Distribution of monoamine nerve terminals in the central nervous system. *Acta Physiol. Scand.* **64**: (Suppl. 247) 37–85.

Fuxe, K. (1965b). Evidence for the existence of monoamine neurons in the central nervous system. 3. The monoamine nerve terminal. *Z. Zellforsch.* **65**:573–596.

Fuxe, K. (1967). Localization and mapping of amine-containing neurons by fluorescent techniques. In Kety, S.S., and Samson, F. E., Jr. (eds.), *Neuroscience Research Program Bulletin,* Vol. 5, pp. 18–22.

Fuxe, K., and Gunne, L. (1964). Depletion of amine stores in brain catecholamine terminals on amygdaloid stimulation. *Acta Physiol. Scand.* **62**:493–494.

Fuxe, K., and Hillarp, N. A. (1964). Uptake of L-dopa and noradrenaline by central catecholamine neurons. *Life Sci.* **3**:1403–1406.

Fuxe, K., and Hökfelt, T. (1966). Further evidence for the existence of tubero-infundibular dopamine neurons. *Acta Physiol. Scand.* **66**:245–246.

Fuxe, K., and Hökfelt, T. (1969). Monoamine afferent input to the hypothalamus and the dopamine afferent input to the median eminence. In: Gaul, C., and Ebling, F. J. G. (eds.), *Progress in Endocrinology,* Amsterdam, pp. 495–502.

Fuxe, K., and Hökfelt, T. (1970). Participation of central monoamine neurons in the regulation of anterior pituitary function with special regard to the neuroendocrine role of tubero-infundibular dopamine neurons. In Bargmann, W., and Scharrer, B. (eds.), *Aspects of Neuroendocrinology,* Springer-Verlag, Berlin, pp. 192–205.

Fuxe, K., and Ljunggren, L. (1965). Cellular localization of monoamines in the upper brain of the pigeon. *J. Comp. Neurol.* **125**:355–381.

Fuxe, K., Hökfelt, T., and Nilsson, O. (1964). Observations on the cellular localization of dopamine in the caudate nucleus of the rat. *Z. Zellforsch.* **63**:701–706.

Fuxe, K., Hökfelt, T., and Nilsson, O. (1965). A fluorescence and electron-miscroscopic study on certain brain regions rich in monoamine terminals. *Am. J. Anat.* **117**:33–45.

Gaddum, J. H. (1953). Antagonism between lysergic acid diethylamide and 5-hydroxytryptamine. *J. Physiol.* (London) **121**:15 p.

Gaddum, J. H., and Giarman, N. J. (1956). Preliminary studies on the biosynthesis of 5-HT. *Brit. J. Pharmacol.* **11**:88–92.

Gaddum, J. H., and Vogt, M. (1956). Some central action of 5-HT and various antagonists. *Brit. J. Pharmacol.* **11**:175–179.

Gal, E. M., and Armstrong, J. C. (1966). The nature of *in vitro* hydroxylation of *l*-tryptophan by brain tissue. *J. Neurochem.* **13**:643–654.

Gal, E. M., and Marshall, F. D. (1965). The hydroxylation of tryptophan by pigeon brain *in vivo.* In Himwich, H. E., and Himwich, W. A. (eds.), *Progr. Brain Res.* **8**:56–60.

Gal, E. M., Morgan, M., Chatterjee, S. K., and Marshall, F. D., Jr. (1964). Hydroxylation of tryptophan by brain tissue *in vivo* and related aspects of 5-hydroxytryptamine metabolism. *Biochem. Pharmacol.* **13**:1639–1653.

Gangloff, H., and Monnier, M. (1955). Local determination of cerebral effects of reserpine. *Experientia* **11**:404–407.

Gangloff, H., and Monnier, M. (1957). Topic action reserpine, serotonin and chlorpromazine on the unanesthetized rabbit's brain. *Helv. Physiol. Acta.* **15**:83–104.

Ganrot, P. O., Rosengren, E., and Gottfries, C. G. (1962). Effect of iproniazid on monoamines and monoamine oxidase in human brain. *Experientia* **18**:260–261.

Ganrot, P. O., Gottfries, C. G., and Rosengren, E. (1963). Effects of some psychotropic drugs on the metabolism of catecholamines and 5-hydroxytryptamine in human brain. *Acta Psychiat. Scand.* **39**: (Suppl. 169) 248.

Garattini, S., and Valzelli, L. (1958). Researches on mechanism of reserpine sedative action. *Science* **128**:1278.

Garattini, S., and Valzelli, L. (1965a). Substances interfering with the narcosis-potentiating activity of serotonin. *Boll. Soc. Ital. Biol. Sper.* **32**:292–295.

Garattini, S., and Valzelli, L. (1965b). *Serotonin,* Elsevier, Amsterdam.

Gascon, A. (1964). Role de quelques libérateurs de la sérotonine et des catécholamines dans le phénomène de la douleur chez le rat. *Rev. Can. Biol.* **23**:169–186.

Gensette, R., Andre-Balisaux, G., and Delmotte, P. (1966). La perméabilité des vaisseaux cérébraux. VI. Démyélinisation expérimentale provoquée par des substances agissant sur la barrière hémotoencéphalique. *Acta Neurol. Belg.* **66**:247–262.

Gerebtzoff, M. A., and Dresse, A. (1961). Contribution à la recherche des catécholamines au miscroscope de fluorescence. *Ann. Histochim.* **6**:125–130.

Gerhards, E., and Gibian, H. (1967). The metabolism of dimethyl sulfoxide and its metabolic effects in man and animals. *Ann N. Y. Acad. Sci.* **141**:65–76.

Gerstenbrand, F. von, Patiesky, K., and Prosenz, P. (1963). Erfahrenheit mit L-Dopa in der Therapie des Parkinsonismus. *Psychiat. Neurol.* **146**:246.

Gessa, G. L., Costa, E., Kuntzman, R., and Brodie, B. B. (1962). Evidence that the loss of brain catecholamine stores due to blockade of storage does not cause sedation. *Life Sci.* **9**: 441–452.

Gey, K. F., and Pletscher, A. (1964). Distribution and metabolism of DL-3,4-dihydroxy [2-^{14}C]-phenylalanine in rat tissue. *Biochem. J.* **92**:300–308.

Giarman, N. J. (1956). Biosynthesis of 5-hydroxytryptamine (serotonin, enteramine). *Fed. Proc.* **15**:428.

Giarman, N. J., and Polter, L. T. (1958). Release of serotonin by extracts of mammalian tissue. *Fed. Proc.* **17**:371.

Giarman, N. J., and Schanberg, S. (1959). The intracellular distribution of 5-HT in the rat's brain. *Biochem. Pharmacol.* **1**:301–306.

Giarman, N. J., Freedman, D. X., and Picard-Ami, L. (1960). Serotonin content of the pineal glands of man and monkey. *Nature* **186**:480.

Giarman, N. J., Green, V. S., Green, J. P., and Paasonen, M. K. (1957). Pharmacological and biochemical study of a carcinoid tumor. *Proc. Soc. Exptl. Biol.* **94**:761.

Gitlow, S. E., Bertani, L. M., Wilk, E., Li, B. L., and Dziedzic, S. (1970). Excretion of catecholamine metabolities by children with familial dysautonomia. *Pediatrics* **46**:513–522.

Gjessing, L. R. (1964). Studies of periodic catatonia. II. The urinary excretion of phenolic amines and acids with and without loads of different drugs. *J. Phychiat. Res.* **2**:149–162.

Gloor, P. (1955). Electrophysiological studies on connections of amygdaloid nucleus in cat. *Electroenceph. Clin. Neurophysiol.* **7**:223–264.

Glowinski, J. (1967). Uptake, storage and metabolism of catecholamines in brain. *Neuroscienes Res. Progr. Bull.* **5**:53

Glowinski, J., and Axelrod, J. (1964). Inhibition of uptake of tritiated-noradrenaline in the intact rat brain by imipramine and structurally related compounds. *Nature* **204**:1318–1319.

Glowinski, J., and Axelrod, J. (1966). Effects of drugs on the disposition of H^3 norepinephrine in the rat brain. *Pharmacol. Rev.* **18**:775–785.

Glowinski, J., and Iversen, L. (1966a). Regional studies of catecholamines in the rat brain. III. Sub-cellular distribution of endogenous and exogenous catecholamines in various brain regions. *Biochem. Pharmacol.* **15**:977–987.

Glowinski, J., and Iversen, L. L. (1966b). Regional studies of catecholamines in the rat brain. I. The disposition of ^{3}H norepininephrine, ^{3}H dopamine, and ^{3}H dopa in various regions of the brain. *J. Neurochem.* **13**:655–669.

Glowinski, J., Kopin, I. J., and Axelrod, J. (1965). Metabolism of H^3 norepinephrine in the rat brain. *J. Neurochem.* **12**:25–30.

Glowinski, J., Iversen, L. L., and Axelrod, J. (1966a). Storage and synthesis of norepinephrine in the reserpine-treated rat brain. *J. Pharmacol. Exptl. Therap.* **151**:385–399.

Glowinski, J., Snyder, S. H., and Axelrod, J. (1966b.) Subcellular localization of H^3-norepinephrine in the rat brain and the effect of drugs. *J. Pharmacol. Exptl. Therap.* **152**:282–292.

Glowinski, J., Axelrod, J., Kopin, I. J., and Wurtman, R. J. (1964). Physiological disposition of H^3-norepinephrine in the developing rat. *J. Pharmacol. Exp. Ther.* **146**:48.

Gluckman, M. I., Hart, E. R., and Marazzi, A. S. (1957). Cerebral synaptic inhibition by serotonin and iproniazid. *Science* **126**:448–449.

Gogerty, J. H., and Horita, A. (1960). A comparison of the *in vivo* inhabition of brain and liver monoamine oxidase as produced by beta-phenylisoprophylhydrazine (PIH) and iproniazid. *J. Pharmacol. Exptl. Therap.* **129**:357–360.

Goldstein, M., and Nakajima, K. (1967). The effect of disulfiram on catecholamine levels in the brain. *J. Pharmacol. Exptl. Therap.* **157**:96–102.

Goldstein, M., Friedhoff, A. J., Simmonds, C., and Prochoroff, N. N. (1959). The metabolism of 3-hydroxytyramine-1-C^{14} in brain tissue homogenates. *Experientia* **15**:254–256.

Goldstein, M., Prochoroff, N., and Sirilin, S. (1965). A radioassay for dopamine-beta-hydroxylase activity. *Experientia* **21**:592–593.

Gomirato, G., and Ferro-Milone, F. (1959). Observations on the action of 5-HT in epileptic patients clinical E. E. G. study. *Electroenceph. Clin. Neurophysiol.* **11**:847.

Goodman, L. S., and Gilman, A. (1955). *The Pharmacological Basis of Therapeutics,* 2 ed., Macmillan, New York, p. 479.

Goodwin, F. K., and Brodie, K. (1970). The combined use of L-dopa and a peripheral decarboxylase inhibitor (MK-485) in depression. Fifty-first Annual American College of Physicians Meeting, Philadelphia, April 12–17, 1970.

Görög, P., and Szporny, L. (1961). Effect of vincamin on the noradrenaline content of rat tissue. *Biochem. Pharmacol.* **8**:259–262.

Grahame-Smith, D. G. (1964). Tryptophan hydroxylation in brain. *Biochem. Biophys. Res. Commun.* **16**:586–592.

Gray, E. G., and Whittaker, V. P. (1960). The isolation of synaptic vesicles from the central nervous system. *J. Physiol. (Lond.)* **153**:35P–39P.

Gray, E. G., and Whittaker, V. P. (1962). The isolation of nerve endings from brain: An electron microscopic study of cell fragments derived by homogenization and centrifugation. *J. Anat.* **96**:79–88.

Grazer, F. M., and Clemente, C. D. (1957). Developing blood–brain barrier to trypan blue. *Proc. Soc. Exptl. Biol.* **94**:758–760.

Green, H., and Erickson, R. W. (1960). Effect of trans-2-phenylcyclopropylamine upon norepinephrine concentration and monoamine oxidase activity of rat brain. *J. Pharmacol. Exptl. Therap.* **129**:237–242.

Green, H., and Sawyer, J. L. (1960). Correlation of tryptamine induced convulsions in rats with brain tryptamine concentration. *Proc. Soc. Exptl. Biol.* **104**:153–154.

Green, H., and Sawyer, J. L. (1964). Biochemical-pharmacological studies with 5-hydroxytryptophan, precursor of serotonin. *Progr. Brain Res.* **8**:153.

Green, H., and Sawyer, J. L. (1966). Demonstration, characterization and assay procedure of tryptophan hydroxylase in rat brain. *Anal. Biochem.* **15**:53.

Green, H., Sawyer, J. L., Erickson, R. W., and Cook, L. (1962). Effect of repeated oral administration of monoamine oxidase inhibitors on rat brain amines. *Proc. Soc. Exptl. Biol.* **109**:347–349.

Griesemer, E. C., Barsky, C., Dragstedt, C., Wells, J. A., and Zeller, E. A. (1953). Potentiating effect of iproniazid on pharmacological actions of sympathomimetic amines. *Proc. Soc. Exptl. Biol.* **84**:699–701.

Gromova, E. A. (1967). Significance of serotonin for central nervous system activity. *Electroenceph. Clin. Neurophysiol.* **22**:585.

Grossman, S. P. (1962). Effects of adrenergic and cholinergic blocking agents on hypothalamic mechanisms. *Am. J. Physiol.* **22**:1230–1236.

Grundfest, H. (1959). General physiology and pharmacology of synapses and some implications for the mammalian central nervous system. *J. Nerv. Ment. Dis.* **128**:473–496.

Gunne, L. M. (1962). Relative adrenaline content in brain tissue. *Acta Physiol. Scand.* **56**:324–333.

Gunne, L. M., and Jonsson, J. (1965). On the occurrence of tyramine in the rabbit brain. *Acta Physiol. Scand.* **64**:434–438.

Gunne, L. M., and Lewander, T. (1966). Monoamines in brain and adrenal glands of cats after electrically induced defense reaction. *Acta Physiol. Scand.* **67**:405.

Gurin, S., and Delluva, A. M. (1947). The biological synthesis of radioactive adrenalin from phenylaline. *J. Biol. Chem.* **170**:545.

Guroff, G., and Udenfriend, S. (1962). Studies on aromatic amino acid uptake by rat brain *in vivo*. *J. Biol. Chem.* **237**:803.

Guroff, G., and Udenfriend, S. (1964). The uptake of aromatic amino acids by the brain of mature and newborn rats. *Progr. Brain Res.* **9**:187–197.

Gursey, D., and Olson, R. E. (1960). Depression of serotonin and norepinephrine levels in brainstem of rabbits by ethanol. *Proc. Soc. Exptl. Biol.* **104**:280–281.

Gyermek, L. (1965). Drugs which antagonize 5-hydroxytryptamine and related indolealkylamines. In Erspamer, V. (ed.), *Handbuch Exptl. Pharmacol.*, Springer-Verlag, Berlin, pp. 471–528.

Haber, B., and Kamano, A. (1966). Sub-cellular distribution of serotonin in the developing rat brain. *Nature* **209**:404.

Haber, B., Khol, H., and Pscheidt, G. R. (1965). Supernatant–particulate distribution of exogenous serotonin in rat brain homogenates. *Biochem. Pharmacol.* **14**:1–6.

Häggendahl, J. (1962). Fluorimetric determinations of 3-O-methylated derivatives of adrenaline and noradrenaline in tissues and body fluids. *Acta Physiol. Scand.* **56**:258–266.

Häggendahl, J. (1967). The effect of high pressure air or oxygen with and without carbon dioxide added on the catecholamine levels of rat brain. *Acta Physiol. Scand.* **69**:147–152.

Häggendahl, J., and Linqvist, M. (1961). Ineffectiveness of ethanol on noradrenaline, dopamine, or 5-hydroxytryptamine levels in brain. *Acta Pharmacol.* **18**:278–280.

Häggendahl, J., and Malmfors, T. (1967). Evidence of dopamine-containing neurons in the retina of rabbits. *Acta Physiol. Scand.* **59**:295–296.

Haley, T. J. (1957). Intracerebral injections of psychotomimetic and psychotherapeutic drugs into conscious cats. *Acta Pharmacol.* **13**:107–112.

Hamberger, A., and Hamberger, B. (1966). Uptake of catecholamines and penetration of trypan blue after blood–brain barrier lesions. A histochemical study. *Z. Zellforsch.* **70**:386–392.

Hamberger, B., and Masuoka, D. (1965). Localization of catecholamine uptake in rat brain slices. *Acta Pharmacol.* **22**:363–368.

Hamberger, B., and Norberg, K. A. (1963). Monoamines in sympathetic ganglia studied with a fluorescence method. *Experientia* **19**:580–583.

Hamberger, B., and Norberg, K. A. (1964). Histochemical demonstration of catecholamines in fresh frozen sections. *J. Histochem. Cytochem.* **12**:48–49.

Hamberger, B., Norberg, K. A., and Sjöqvist, S. (1963). Cellular localization of monoamines in sympathetic ganglia of the cat. *Life Sci.* **9**:659–661.

Hamberger, B., Norberg, K. A., and Sjöqvist, S. (1964). Evidence for adrenergic nerve terminals and synapses in sympathetic ganglia. *Internat. J. Neuropharmacol.* **2**:279–282.

Hamberger, B., Malmfors, T., and Sachs, C. (1965). Standardization of paraformaldehyde and of certain procedures for histochemical demonstration of catecholamines. *J. Histochem. Cytochem.* **13**:147.

Hamberger, B., Ritzén, M., and Wersäll, J. (1966). Demonstration of catecholamines and 5-hydroxytryptamine in the human carotid body. *J. Pharmacol. Exptl. Therap.* **152**:197–201.

Hammond, J. B. (1956). The effect of serotonin and LSD on the secretory response to reserpine. *Clin. Res. Proc.* **4**:247.

Hanna, C. (1965). Metabolism of catecholamines. *Invest. Ophthalmol.* **4**:1095–1104.

Hanson, A. (1966). Chemical analysis of indolalkylamines and related compounds. In Erspamer, V. (ed.), *Handbuch Exptl. Pharmacol.*, Springer-Verlag, Berlin, pp. 66–112.

Hanson, L. C. (1965). The disruption of conditioned avoidance response following selective depletion of brain catecholamines. *Psychopharmacologia* **8**:100–110.

Hansson, E. (1963). Effect of nicotine on the catecholamine and the serotonin level in some tissues. *Proc. West. Pharmacol. Soc.* **6**:36–37.

Harrison, T. L. (1966). Tissue content of epinephrine and norepinephrine following adrenal medullectomy. *Am. J. Physiol.* **210**:599–600.

Hartmann, E. (1968). On the pharmacology of dreaming sleep. *J. Nerv. Ment. Dis.* **146**:165.

Harvey, J. A. (1965). Comparison between the effects of hypothalamic lesions on brain amine levels and drug action. *J. Pharmacol. Exptl. Therap.* **147**:244–251.

Harvey, J. A., Heller, A., and Moore, R. Y. (1963). The effect of unilateral and bilateral medial forebrain bundle lesions on brain serotonin. *J. Pharmacol. Exptl. Therap.* **140**:103–110.

Hashimoto, Y., Ishii, S., Chi, Y., Shimizu, N., and Imaizumi, R. (1965). Effects of dopa on the norepinephrine and dopamine contents and on the granulated vesicles of the hypothalamus of reserpinized rabbits. *Jap. J. Pharmacol.* **15**:395–400.

Haškovec, L. and Souček, K. (1968). Trial of methysergide in mania. *Nature* **219**:507–508.

Haverback, B. J., Shore, P. A., Tomich, E., and Brodie, B. B. (1956). Cumlative effect of small doses of reserpine on serotonin in man. *Fed. Proc.* **15**:434–435.

Heath, R. G. (1964). Pleasure response of human subjects to direct stimulation of the brain: physiologic and psychodynamic considerations. In *The Role of Pleasure in Behavior*, R. G. Heath (ed.), New York, Hoeber-Harper and Row, 219–243.

Heath, R. G. (1971. Depth recording and stimulation studies in patients. In Winter, A. (ed.), *The Surgical Control of Behavior*, Charles C. Thomas, Springfield, 21–37.

Heath, R. G., and Krupp, I. M. (1967). Schizophrenia as an immunological disorder. *Arch. Gen. Psychiat.* **16**:1–33.

Heller, A., and Moore, R. Y. (1970). Localization and neural control of brain monoamines. In Himwich, H. E. (ed.), *Biochemistry, Schizophrenia and Affective Illnesses*, Williams and Wilkins Co., Baltimore.

Heller, A., Harvey, J. A., and Moore, R. Y. (1962). A demonstration of a fall in brain serotonin following central nervous system lesions in the rat. *Biochem. Pharmacol.* **11**:859–866.

Heller, A., Seiden, L. S., and Moore, R. Y. (1966). Regional effects of lateral hypothalamic lesions on 5-hydroxytryptophan decarboxylase in the cat brain. *J. Neurochem.* **13**:967–974.

Heller, B., Narasimhachari, N., Spaide, J. Haskovec, L., and Himwich, H. E. (1970). *N*-dimethylated indoleamines in blood of acute schizophrenics. *Experientia* **15**:503–504.

Hemmingsen, A. M. (1932). Studies on the oestrus producing hormone (oestrin). *Scand. Arch. Psysiol.* **65**:97–250.

Hempel, K. (1965). Zur Fixierung von radioaktivmarkierten Catecholaminen bei autoradiographischen Untersuchungen. *Histochemie* **4**:507–513.

Henning, M. (1969). Interaction of dopa decarboxylase inhibitors with the effect of alpha-methyldopa on blood pressure and tissue monoamines in rats. *Acta Pharmac. Toxicol.* **27**:135–148.

Henning, M., and Rubenson, A. (1970). Central hypotensive effect of L-3,4-dihydroxyphenylalamine in the rat. *J. Pharm. Pharmac.* **22**:553–560.

Herberg, L. J., and Blundell, J. E. (1967). Lateral hypothalamus: Hoarding behavior elicited by electrical stimulation. *Science* **155**:349–350.

Hertting, G., and Axelrod, J. (1961). Fate of tritiated noradrenaline at sympathetic nerve endings. *Nature* **192**:172–173.

Hess, S. M., and Doepfner, W. (1961). Behavioral effects and brain amine content in rats. *Arch. Int. Pharmacodyn.* **134**:89–99.

Hess, S. M., Shore, P. A., and Brodie, B. B. (1956a). Persistence of reserpine action after its disappearance from brain. *Fed. Proc.* **15**:437.

Hess, S. M., Shore, P. A., and Brodie, B. B. (1956b). Persistence of reserpine action after the disappearance of drug from brain-effect on serotonin. *J. Pharmacol. Exptl. Therap.* **118**:84.

Hess, S. M., Ozaki, S., and Udenfriend, S. (1960). The effects of α-methyl dopa and α-methyl metatyrosine in the metabolism of serotonin and norepinephrine. *Pharmacologist* **2**:81.

Hess, S. M., Connamacher, R. H., Ozaki, M., and Udenfriend, A. (1961). The effect of α-methyl dopa and α-methyl-meta-tyrosine on the metabolism of norepinephrine and serotonin *in vivo*. *Arch. Int. Pharmacodyn.* **134**:129–138.

Hess, W. R. (1936). Hypothalamus und die Zentren des autonomen Nervensystems. *Arch. Psychiatr. u. Nervenkrankh.* **104**:548.

Hess, W. R. (1956). *Hypothalamus und Thalamus*, Experimental-Dokumente, Vol. 1, Thieme, Stuttgart.

Hess, W. R. (1964). *The Biology of the Mind*, University of Chicago Press, Chicago

Hess, W. R. (1965). Sleep as a phenomenon of the integral organism. *Progr. Brain Res.* **18**: 3–8.

Heymans, C. (1965). Brain amines and electroshock convulsive threshold. *Indian J. Physiol. Pharmacol.* **8**:113–116.

Higuchi, H. (1962). Effect of intravenous and intracarotid administration of reserpine on catecholamine contents of the brain, heart and adrenal gland in rabbit. *Jap. J. Pharmacol.* **12**:34–47.

Hill, R. T., Koosis, I., Minor, M. W., and Sigg, E. G. (1961). Potentiation of methylphenidate by imipramine amitriptyline and their desmethyl analogues. *Pharmacologist* **3**:75.

Hillarp, N. A., Fuxe, K., and Dahlström, A. (1966). Demonstration and mapping of central neurons containing dopamine, noradrenaline, and 5-hydroxytryptamine and their reaction to psychopharmaca. *Pharmacol. Rev.* **18**:727–741.

Himwich, H. E., and Rinaldi, F. (1957). *Brain Mechanisms and Drug Action*, Charles C. Thomas, Springfield, Ill., pp. 15–44.

Hoagland, H. (1957). A review of biochemical changes *in vivo* by lysergic acid diethylamide and similar drugs. *Ann N. Y. Acad. Sci.* **66**:445–458.

Hoffer, A. (1957). Epinephrine derivatives as potential schizophrenic factors. *J. Clin. Psychopathol.* **18**:27–60.

Hoffer, A., and Osmond, H. (1959). The adrenochrome and schizophrenia. *J. Nerv. Ment. Dis.* **128**:18–32.

Hoffer, A., Osmond, H., Callbeck, M. J., and Kahan, I. (1957). Treatment of schizophrenia with nicotinic acid and nicotinamide. *J. Clin. Exptl. Psychopathol.* **18**:131–158.

Hofmann, A. (1959). Psychotomimetic drugs, chemical and pharmacological aspects. *Acta Physiol. Pharmacol.* **8**:240–258.

Hökfelt, T. (1965). A modification of the histochemical fluorescence method for the demonstration of catecholoamines and 5-hydroxytryptamine, using Araldite as embedding medium. *J. Histochem. Cytochem.* **13**:518–520.

Hökfelt, T. (1967). Electron microscopic studies on brain slices from regions rich in catecholamine nerve terminals. *Acta Physiol. Scand.* **69**:119–120.

Hollister, L. E. (1964). Complications from psychotherapeutic drugs. *Clin. Pharmacol. Ther.* **5**:322.

Holtz, P. (1950). Uber die sympathicomimetische Wirksamkeit von Gehirnextrakten. *Acta Physiol. Scand.* **20**:354–362.

Holzbauer, M., and Vogt, M. (1956). Depression by reserpine of the noradrenaline concentration in the hypothalamus of the cat. *J. Neurochem.* **1**:8–11.

Holzer, G., and Hornykiewicz, O. (1959). On dopamine metabolism in the rat brain. *Arch. Exptl. Pathol.* **237**:27–33.

Hopkin, D. A. B., and Brown, D. (1958). The reticular system and chlorpromazine. *Anesthesia* **13**:454–456.

Horden, A. (1965). The antidepressant drugs. *New Engl. J. Med.* **272**:1159–1169.

Horita, A., and McGrath, W. R. (1960). The interaction between reversible and irreversible monoamine oxidase inhibitors. *Biochem. Pharmacol.* **3**:206.

Hornykiewicz, O. (1963). Die topische Lokalisation und das Verhalten von Noradrenalin und Dopamine (3-Hydroxytryramin) in der Substantia nigra des Normalen und parkinsonkranken Menschen. *Wien. Klin. Wschr.* **75**:309–312.

Hornykiewicz, O. (1964). The role of dopamine (3-hydroxytyramine) in parkinsonism. In

Trabucchi, E., Paoletti, R., and Canal, N. (eds.), *Biochemical and Neurophysiological Correlations of Centrally Acting Drugs*, Pergamon Press, Oxford.

Hornykiewicz, O. (1966). Dopamine (3-hydroxytyramine) and brain function. *Pharmacol. Rev.* **18**:925–964.

Hsia, D. Y., Nishimura, K., and Brenchley, Y. (1963). Mechanisms for the decrease of brain serotonin. *Nature* **200**:578 *ff.*

Hsia, D. Y., Justice, P., Berman, J. L., and Brenchley, Y. (1964). Brain serotonin in experimental tyrosinosis. *Nature* **202**:495–496.

Hughes, F., Shore, P. A., and Brodie, B. B. (1958). Serotonin storage mechanism and its interaction with reserpine. *Experientia* **14**:178–179.

Huttunen, M. O. (1971). Persistent alteration of turnover of brain noradrenaline in the offspring of rats subjected to stress during pregnancy. *Nature* **230**:53–55.

Ikeda, M., Fahien, L. A., and Udenfriend, S. (1966). A kinetic study of bovine adrenal tyrosine hydroxylase. *J. Biol. Chem.* **241**:4452–4456.

Ingram, W. R. (1940). Nuclear organization and chief connections of the primate hypothalamus. *Proc. Assoc. Res. Nerv. Ment. Dis.* **20**:195–244.

Inscoe, J. K., Daly, J., and Axelrod, J. (1965). Factors affecting the enzymatic formation of *O*-methylated dihyroxy derivatives. *Biochem. Pharmacol.* **14**:1257–1263.

Isbell, H., Logan, C. R., and Miner, E. J. (1959). Studies on LSD 25. Attempts to attenuate the LSD reaction in a man by pre-treatment with neurohumoral blocking agents. *Arch. Neurol. Psychiat.* **81**:20–27.

Ishii, S., Shimizu, N., Matsuoka, M., and Imaizumi, R. (1965). Correlation between catecholamine content and numbers of granulated vesicles in rabbit hypothalamus. *Biochem. Pharmacol.* **14**:183–184.

Ishikawa, K., Ott, K., and Eldred, E. (1967). Response of motoneurons to hypothalamic stimulation and its relation to thermoregulatory behavior. *Am. J. Phys. Med.* **46**:1290–1301.

Ishikawa, T., Koizumi, K., and Brooks, C. (1966). Activity of supraoptic nucleus neurons of the hypothalamus. *Neurology* **16**:101–106.

Itoh, T., Matsuoka, M., Nakazima, K., Tagawa, K., and Imaizumi, R. (1962). An isolation method of catecholamine and effect of reserpine on the enzyme systems related to the formation and inactivation of catecholamines in brain. *Jap. J. Pharmacol.* **12**:130–136.

Itoh, T., Kajikawa, K., Hashimoto, Y., Yoshida, H., and Imaizumi, R. (1965). The uptake of catecholamines by subcellular granules in the brain stem. *Jap. J. Pharmacol.* **15**:335–338.

Iversen, L. L. (1966). Regional studies of catecholamines in the rat brain. I. Rate of turnover of catecholamines in various brain regions. *J. Neurochem.* **13**:671–682.

Iversen, L. L. (1967). *The Uptake and Storage of Noradrenaline in Sympathetic Nerves*, Cambridge University Press.

Iwamoto, T., and Satoh, T. (1963). Effects of chlorpromazine, azacyclonol, and chlordiazepoxide on brain catecholamine contents of stressed rats. *Jap. J. Pharmacol.* **13**:66–73.

Izquierdo, J. A., Jofre, I. J., and Dezza, M. A. (1964). Effect of pyrogallol on the catecholamine contents of cortex, diencephalon, mesencephalon, rhomboencephalon, and cerebellum of mouse and rat. *Med. Exp.* **10**:45–55.

Jacobowitz, D. (1965). A method for the demonstration of both acetylcholinesterase and catecholamine in the same nerve trunk. *Life Sci.* **4**:297–303.

Jacobowitz, D., and Koelle, G. B. (1965). Histochemical correlations of acetylcholinesterase and catecholamines in postganglionic autonomic nerves of the cat, rabbit and guinea pig. *J. Pharmacol. Exptl. Therap.* **148**:225–237.

Jepson, J. B. (1958). *Chromatographic Techniques*, William Heinemann, London, p. 114.

Johannesson, T., and Lausen, H. H. (1961). Chlorpromazine as an inhibitor of brain cholinesterases. *Acta Pharmacol.* **18**:398–406.

Johansson, B., Li, C.-L., Olsson, Y., and Klatzo, I. (1970). The effect of acute arterial hypertension on the blood-brain barrier to protein tracers. *Acta Neuropathol.* **16**:117–124.

Johnson, G. A., Boukma, S. J., and Kim, E. G. (1970). *In vivo* inhibition of dopamine-β-hydroxylase by 1-phenyl-3-(2-thiazolyl)-2-thiourea (U-14, 624). *J. Pharmacol. Exptl. Therap.* **171**:80–87.

Johnson, G. E. (1964). The influence of chlorpromazine on the catecholamine excretion of normal and cold acclimatized rats. *Acta Physiol. Scand.* **60**:181–188.

Johnson, H. A., Smith, R. E., and Simon, W. (1958). Diurnal variations and the excretion of 5-hydroxyindoleacetic acid. *Clin. Res.* **6**:268.

Johnson, T. N. (1965). An experimental study of the fornix and hypothalamotegmental tracts in the cat. *J. Comp. Neurol.* **125**:29–39.

Johnston, J. B. (1923). Further contributions to the study of the evolution of the forebrain. *J. Comp. Neurol.* **35**:337–442.

Jonason, J., Rosengren, E., and Waldeck, B. (1963). Effects of some pharmacologically active amines on the uptake of arylkylamines by adrenal medullary granules. *Acta Physiol. Scand.* **60**:136–140.

Jonec, V. (1966). Interaction of stress reactions in the pituitary–adrenocortical system during telestimulation of the hypothalamus in rats. *Bratisl. Lek. Listy.* **46**:47–52.

Joo, F., and Osillik, B. (1966). Topographic correlation between the hemato-encephalic barrier and the cholinesterase activity of brain capillaries. *Exptl. Brain Res.* **1**:147.

Jouvet, M. (1967a). Neurophysiology of the states of sleep. *Physiol. Rev.* **47**:117–177.

Jouvet, M. (1967b). Mechanisms of the states of sleep. A neuropharmacological approach in sleep and altered states of consciousness. *Ass. Res. Nerv. Dis. Proc.,* Williams and Wilkins Co., Baltimore.

Jouvet, M. (1968). Insomnia and decrease of the cerebral 5-hydroxytryptamine after destruction of the raphé system in the cat. In Garattini, S., and Shore, P. A. (eds.), *Advances in Pharmacology,* Academic Press, New York, pp. 265–279.

Jouvet, M., and Delorme, F. (1965). Locus coeruleus et sommeil paradoxal. *Compt. Rend. Soc. Biol. (Paris)* **159**:895–899.

Jouvet, M., and Renault, J. (1966). Insomnie persistante après lesions des noyaux du raphé chez le chat. *Compt. Rend. Soc. Biol. (Paris)* **160**:1461.

Jouvet, M., Vimont, P., and Delorme, F. (1965). Suppression élective du sommeil paradoxal chez le chat par les inhibiteurs de la monoamineoxydase. *Compt. Rend. Soc. Biol. (Paris)* **159**:1595–1599.

Jouvet, M., Bobillier, P., Pujol, J. F., and Renault, J. (1967). Suppression du sommeil et diminution de la sérotonine cérébrale par lesion du système du raphé chez le chat. *Compt. Rend. Acad. Sci. (Paris)* **264**:360–367.

Justice, P., and Hsia, D. Y. (1965). Studies on inhibition of brain 5-hydroxytryptophan decarboxylase by phenylalanine metabolites. *Proc. Soc. Exptl. Biol.* **118**:326–328.

Kabat, H., Magoun, H. W., and Ranson, S. W. (1934). Electrical stimulation of the hypothalamus. *Proc. Soc. Exptl. Biol.* **31**:541.

Kajikawa, H. (1969). Mode of the sympathetic innervation of the cerebral vessels demonstrated by the fluorescent histochemical technique in rats and cats. *Arch. Jap. Chir.* **38**:227–235.

Kakimoto, Y. (1964). Determination of amines in brain tissue. I. Measurement of amines in the brain. *Brain Nerve* **16**:398–401.

Kaliuzhnyi, L. V. (1962). Changes in food and defensive conditioned reflexes under the influence of the administration of noradrenaline and carbocholine into the posterior hypothalamus. *Zh. Vyssh. Nerv. Dey tavlova* **12**:318–325.

Kamberi, I. A., and Kobayashi, Y. (1970). Monoamine oxidase activity in the hypothalamus and various other brain areas and in some endocrine organs of the rat during the estrus cycle. *J. Neurochem.* **17**:261–268.

Kamberi, I. A., and McCann, S. M. (1969). Effect of biogenic amines, FSH-releasing factor (FRF) and other substances on the release of FSH by pituitaries incubated *in vitro*. *Encocrinology* **85**:815–824.

Kappers, J. A. (1960). The development, topographical relations and innervation of the epiphysis cerebri in the albino rat. *Z. Zellforsch. Microskop. Anat.* **52**:163.

Karki, N., Kuntzman, R., and Brodie, B. B. (1962). Storage, synthesis and metabolism of monoamines in the developing brain. *J. Neurochem.* **9**:53–58.

Karlsson, P. (1965). Biologic oxidation—metabolism of oxygen. In *Introduction to Modern Biochemistry*, Academic Press, New York, p. 193.

Kaskel, D. (1965). Morphologische Beobachtungen im Hypothalamus von adrenalektomierten Hunden und Katzen. *Acta Anat.* **62**:343–364.

Katz, R. I., Chase, T. N., and Kopin, I. J. (1968). Evoked release of norepinephrine and serotonin from brain slices: Inhibition by lithium. *Science* **162**:466.

Kawamura, H., and Nakamura, Y. (1961). Role of the hypothalamus in the brain activating system. *Brain Nerve* **13**:857–861.

Kayser, Ch. (1961). *The Physiology of Natural Hibernation.* Pergamon Press, New York, p. 286–287.

Kesarev, V. S. (1966). Structural features of hypothalamus of man and other primates (chimpanzee, macaque). *Fed. Proc.* **25**: (Trans. Suppl.) 243–247.

Kety, S. S. (1966). Catecholamines in neuropyschiatric states. *Pharmacol. Rev.* **18**:787–798.

Kety, S. S. (1970). The biogenic amines in the central nervous system: Their possible roles in arousal, emotion and learning. In Schmitt, F. O. (ed.), *The Neurosciences: Second Study Program*, Rockefeller Press, pp. 324–335.

Khalifeh, R. R., Kaelber, W. W., and Ingram, W. (1965). Some efferent connections of the nucleus medialis dorsalis. *Am. J. Anat.* **116**:341–354.

Kimishima, K. (1963). Effects of reserpine and biogenic amines on hippocampal seizure discharges. *Yenago Acta Med.* **7**:1–6.

King, F. A. (1958). Effects of septal and amygdaloid lesions on emotional behavior and conditioned avoidance response in the cat. *J. Nerv. Ment. Dis.* **126**:57.

Kirshner, N. (1957). Pathway of noradrenaline formation from dopa. *J. Biol. Chem.* **226**:821.

Kirshner, N. (1962). Uptake of catecholamines by particulate fraction of adrenal medulla. *J. Biol. Chem.* **237**:2311–2317.

Kirshner, N. (1966). The function of catecholamines in the brain. *J. Neurosurg.* **24**:165–167.

Kissel, J. W., and Domino, E. F. (1959). The effects of some possible neurohumoral agents on spinal cord reflexes. *J. Pharmacol. Exptl. Therap.* **125**:168–177.

Kivalo, E., and Rinne, U. K. (1959). Effect of 5-hydroxytryptamine on the nucleus supraopticus. *Ann. Med. Exp. Fenn.* **37**:262–268.

Kivalo, E., Rinne, U. K., and Karinkanta, H. (1961). The effect of imipramine on the 5-hydroxytryptamine content and monoamine oxidase activity of the rat brain and on the excretion of 5-hydroxyindole acetic acid. *J. Neurochem.* **8**:105–108.

Klaue, R. (1937). Die bioelektrische Tätigkeit der Grosshirnrinde im normalen Schlaf und in der Narkose durch Schlafmittel. *J. f. Psychol. u. Neurol.* **47**:510.

Klee, J. D., Vertino, J., Callaway, E. and Weintraub, W. (1960). Clinical studies with LSD and two substances related to serotonin. *J. Ment. Sci.* **106**:301–308.

Klerman, G. L., and Cole, J. O. (1965). Clinical pharmacology of imipramine and related antidepressant compounds. *Pharmacol. Rev.* **17**:101–141.

Knox, W. E. (1960). Evaluation of the treatment of phenylketonuria with diets low in phenylalamine. *Pediatrics* **26**:1.

Kobayashi, T. (1964). Fluctuations in monoamine oxidase activity in the hypothalamus of rat during the estrous cycle and after castration. *Endocrinol. Jap.* **11**:283–290.

Koe, B. K., and Weissman, A. (1966a). Marked depletion of brain serotonin by *p*-chlorophenylalanine. *Fed. Proc.* **25**:452.

Koe, B. K., and Weissman, (1966b). *p*-Chlorophenylalanine: A specific depletor of brain serotonin. *J. Pharmacol. Exptl. Therap.* **154**:499–516.

Koella, W. P. (1962). Zum Wirkungsmechanismus von Serotonin auf das Zentralnervensystem. *Praxis* **51**:1267–1274.

Koella, W. P., Smythies, J. R., and Bull, D. M. (1959). Factors involved in the effect of serotonin on the evoked electrocortical potentials. *Science* 129:1231.

Koelle, G. B. (1958). Pharmacologic significance of inhibition of monoamine oxidase. *J. Clin. Exptl. Psychopathol.* 19: (Suppl.) 2.

Koelle, G. B., and Valk, A. (1954). Physiological implications of the histochemical localization of the monoamine oxidase. *J. Physiol. (Lond.)* 126:434–447.

Konig, J. F. R., and Klippel, R. A. (1963). *The Rat Brain. A Stereotaxic Atlas of the Forebrain and Lower Parts of the Brain Stem,* Waverly Press, Inc., Baltimore.

Konzett, H. (1956). The effects of 5-HT and its antagonists on tidal air. *Brit. J. Pharmacol.* 11: 289–294.

Kopin, I. J. (1960). Technic for the study of alternate metabolic pathway epinephrine metabolism in man. *Science* 131:1372–1374.

Kopin, I. J. (1964). Storage and metabolism of catecholamines: The role of monoamine oxidase. *Pharmacol. Rev.* 16:179–191.

Kopin, I. J., and Axelrod, J. (1963). The role of monoamine oxidase in the release and metabolism of epinephrine, *Ann. N. Y. Acad. Sci.* 107:848–853.

Kopin, I. J., Fischer, J. F., Musacchio, J., and Horst, W. D. (1964). Induction of false neurochemical transmitters as a mechanism for the sympathetic blocking action of monoamine oxidase inhibitors (MAOI). *Pharmacologist* 6:175.

Krajl, M. (1965). A rapid microfluorimetric determination of monoamine oxidase. *Biochem. Pharmacol.* 14:1684–1686.

Krieg, W. J. (1932). Hypothalamus of albino rat. *J. Comp. Neurol.* 55:19–85.

Kuehl, F. A., Hitchens, M., Ormond, R. E., Meisinger, M. A. P., Gale, P. H., Cirillo, V. J., and Brink, N. G. (1964). Para O-methylation of dopamine in schizophrenic and normal individuals. *Nature* 203:154.

Kuntzman, R., Udenfriend, S., Tomich, E. G., Brodie, B. B., and Shore, P. A. (1956). Biochemical effect of reserpine on serotonin binding sites. *Fed. Proc.* 15:450.

Kuntzman, R., Mead., J. R., Brodie, B. B., and Shore, P. A. (1958). Comparative metabolism of brain serotonin and norepinephrine *in vivo. J. Pharmacol.* 122:41A.

Kuntzman, R., Costa, E., Gessa, G. L., and Brodie, B. B. (1962). Reserpine and guanethidine action on peripheral stores of catecholamines. *Life Sci.* 3:65–74.

Kurachi, K. (1965). Relational function of the hypothalamus, pituitary gland and gonads. *Clin. Endocrinol.* 13:730–738.

Kurachi, K., and Hirota, K. (1969). Catecholamine metabolism in rat's brain related with sexual cycle. *Endocr. Jap. Suppl.* 1:69.

Kuruma, I. (1966). Changes in the concentration of serotonin and catecholamines in febrile rabbits. 2. Effect of high ambient temperature on serotonin and noradrenaline. *Folia Pharmacol.* 62:8–10.

Laasberg, L. H., and Shimosato, S. (1966). Paper chromatographic identification of catecholamines. *J. Appl. Physiol.* 21:1929–1934.

LaBrosse, E. H., Axelrod, J., and Kety, S. (1958). O-methylation, the principal route of metabolism of epinephrine in man. *Science* 128:593–594.

Lagunoff, D., Phillips, M., and Benditt, E. P. (1961). The histochemical demonstration of histamine in mast cells. *J. Histochem.* 9:534–541.

Lajtha, A. (1958). The biochemistry of hallucinogens. *Progr. Neurobiol.* 3:126–151.

Langemann, H., and Goerre, J. (1957). Noradrenaline and its significance in pharmacology. *Schweiz. med. Wochschr.* 87:607.

Laparra, J. (1965). Action of catecholamines on cytochrome *c. Compt. Rend. Acad. Sci. (Paris)* 261:4897–4900.

Laverty, R., and Sharman, D. F. (1965). Modification by drugs of the metabolism of 3,4-dihyroxyphenylethylamine, noradrenaline, and 5-hydroxytryptamine in the brain. *Brit. J. Pharmacol.* 24:759–772.

Leblanc, J., and Pouliot, M. (1964). Importance of noradrenaline in cold adaptation. *Am. J. Physiol.* 207:853–856.

Lefranc, G. (1964). Etude neurohistologique de l'hypothalamus du cobaye par la technique de triple imprégnation de Golgi. 1. La substance grise périventriculaire. *Compt. Rend. Soc. Biol. (Paris)* **158**:2087–2089.

Leme, J. G., Rocha, E., and Silva, M. (1963). Effect upon the analgestic action of reserpine of central nervous system stimulants and drugs affecting the metabolism of catechol- and indole amine. *J. Pharm. Pharmacol.* **15**:454–460.

Lenz, H. (1966). The significance of catecholamines in the clinical picture of schizophrenia. *Wien. Z. Nervenheilk.* **24**:105–109.

Leonardelli, J. (1969). Catecholamines hypothalamiques et contrôle du cycle oestral chez le cobaye et le rat. *Ann. Endocrinol.* **30**:783–788.

Leusen, I. (1960). Monoamin-Freisetzer und Monoaminoxydase-Hemmer. *Psychiat. Neurol.* **140**:154–164.

Leusine, E., Lacroix, E., and Demeester, G. (1959). Quelques propriétés pharmacodynamiques de la tetrabenzine, substance liberatrice de serotonine. *Arch. Int. Pharmacodyn.* **119**:225–231.

Lever, J. D., Lewis, P. R., and Boyd, J. D. (1959). Observations of the structure and histochemistry of carotid body in cat and rabbit. *J. Anat.* **93**:478–490.

Levin, E. Y., and Kaufman, S. (1961). Studies on the enzyme catalyzing the conversion of 3,4-dihydroxyphenylalanin to norepinephrine. *J. Biol. Chem.* **236**:2043.

Levine, R. J., and Sjöerdsma, A. (1963a). A direct measurement of monoamine oxidase inhibition in humans. *Clin. Pharmacol. Therap.* **4**:22–27.

Levine, R. J., and Sjöerdsma, A. (1963b). Estimation of monoamine oxidase activity in man: Techniques and applications. *Ann. N. Y. Acad. Sci.* **107**:848–855.

Lichardus, B. (1965). Size of cell nuclei in the hypothalamus of the rat as a function of salt loading. *Am. J. Physiol.* **208**:1075–1077.

Lichtensteiger, W., and Langemann, H. (1966). Uptake of exogenous catecholamines by catecholamine-containing neurons of the central nervous system: Uptake of catecholamines by arcuate–infundibular neurons. *J. Pharmacol. Exptl. Therap.* **151**:400–408.

Lindmar, R., and Muscholl, E. (1964). Die Wirkung von Pharmaka auf die Elimination von Noradrenalin aus der Perfusionsflüssigkeit und die Noradrenalinaufnahme in das isolierte Herz. *Arch. Exptl. Pathol. Pharmakol.* **247**:469–492.

Lippman, W., and Wishnick, M. (1965). Effect of DL-*p*-chloro-*N*-methylamphetamine on the concentrations of monoamines in the cat and rat brain and rat heart. *Life Sci.* **4**:849–857.

Lippman, W., Leonardi, R., and Ball, J. (1967). Relationship between catecholamines and gonadotropin secretion in rats. *J. Pharmacol. Exptl. Therap.* **156**:258–266.

Lisk, R. D. (1962). Diencephalic placement of estradiol and sexual receptivity in the female rat. *Am. J. Physiol.* **203**:493–496.

Löfving, B. (1961). Cardiovascular adjustments induced from rostral cingulate gyrus. *Acta Physiol. Scand.* **53**: Suppl. 184.

Losovski, D. B. (1965). Biochemical theories of schizophrenia. *Internat. J. Psychiat.* **1**:438–422.

Lovenberg, W., Weissbach, H., and Udenfriend, S. (1962). Aromatic 1-aminoacid decarboxylase. *J. Biol. Chem.* **237**:89.

Lovenberg, W., Jequier, E., and Sjöerdsma, A. (1968). Tryptophan hydroxylation in mammalian systems. In Garattini, S., and Shore, P. A. (eds.), *Advances in Pharmacology*. **6A**:21–36.

Lycke, E., Modigh, K., and Roos, B. E. (1969). Aggression in mice associated with changes in the monoamine-metabolism of the brain. *Experientia* **25**:951–953.

Maas, J. W., and Colburn, R. W. (1965). Co-ordination chemistry and membrane function with particular reference to the synapse and catecholamine transport. *Nature* **208**:41–46.

Maas, A. R., and Nimme, M. J. (1959). A new inhibitor of serotonin metabolism. *Nature* **184**:547–548.

Maccari, M., and Maggi, G. C. (1965). Effects of gamma-amino-beta-hydroxybutyric acid on cerebral amines. *Brit. J. Pharmacol.* 24:462–465.

Maclean, R., Nicholson, W. J., Pare, C. M. B., and Stacey, R. S. (1965). Effect of monoamine-oxidase inhibitors on the concentrations of 5-hydroxytryptamine in the human brain. *Lancet* 2:205–208.

Mairova, V. F. (1962). Change in the neurosecretion of the hypothalamus following hypophy-sectomy. *Probl. Endokr. Germonoter* 8:21–25.

Maitre, L. (1965). Presence of alpha-methyl-DOPA metabolities in heart and brain of guinea pigs treated with alpha-methyl-tyrosine. *Life Sci.* 4:2249–2256.

Malkotra, C. L., and Prasad, K. (1962). Effect of chlorpromazine and reserpine on the cate-cholamine content of different areas of the central nervous system of the dog. *Brit. J. Pharmacol.* 18:595–599.

Malmejac, J., and Chardon, G. (1959). Action de l'adrénaline sur l'activité du cortex cérébral à basse température. *Bull. Acad. Anat. Med.* 143:271–276.

Malmfors, T. (1963). Evidence of adrenergic neurons with synaptic terminals in the retina of rats demonstrated with fluorescence and electron miscroscopy. *Acta Physiol. Scand.* 58:99–100.

Malmfors, T. (1965). Studies on adrenergic nerves. *Acta Physiol. Scand.* 64: (Suppl. 248) 1–93.

Manger, W. H., Schuarz, B. E., Beard, C. W., Wakem, K. G., Bellman, J. L., Peterson, M., and Berkson, J. (1957). Epinephrine and norepinephrine in mental disease. *Arch. Neurol. Psychiat.* 79:396–412.

Mannarino, E., Kirshner, N., and Nashold, B. (1962). The metabolism of noradrenaline-C^{14} by cat brain *in vivo*. *Fed. Proc.* 21:182.

Mantegazzini, P., Poeck, J., and Santibanez, G. (1959). Effect of adrenaline and noradrenaline on the cortical electrical activity of the brain isolated cat. *Pflügers Arch. Ges. Physiol.* 270:14–15.

Marañon, G. (1924). Contribution à l'étude de l'action émotive de l'adrénaline. *Rev. Franc. Endocrinol.* 2:301.

Marchbanks, R. M., Rosenblatt, F., and O'Brian, R. D. (1964). Serotonin binding to nerve-ending particles of the rat brain and its inhibition by lysergic acid diethylamide. *Science* 144: 1135–1137.

Mardsen, D. (1965). Brain pigment and its relation to brain catecholamines. *Lancet* 2:475–476.

Markley, E. (1966). Behavioral and electrophysiological effects of catecholamines. *Pharmacol. Rev.* 18:753–768.

Marrazzi, A. S. (1957). The effects of certain drugs on cerebral synapses. *Ann. N. Y. Acad. Sci.* 66:496–507.

Masai, H., Kusunoki, T., and Ishibashi, H. (1965). A histochemical study on the fundamental plan of the central nervous system. *Experientia* 21:572–573.

Masuoka, D. T., Schott, H. F., and Petriello, L. (1963). Formation of catecholamines by various areas of cat brain. *J. Pharmacol. Exptl. Therap.* 139:73–76.

Matlina, E. (1964). Current views in the synthesis and conversion of catecholamines in the organism. *Usp. Severem. Biol.* 58:321–345.

Matsui, T., and Kobayashi, H. (1965). Histochemical demonstration of monoamine oxydase the hypothalamo-hypophyseal system of the tree sparrow and cat. *Z. Zellforsch.* 68:172–182.

Matsumoto, J., and Jouvet, M. (1964). Effects de réserpine, dopa et 5-HTP sur les deux états de sommeil. *C. R. Soc. Biol.* 158:130.

Matsuoka, M., Yoshida, H., and Imaizumi, R. (1964). Distribution of catecholamines and their metabolites in rabbit brain. *Biochim. Biophys. Acta* 82:429–441.

Matsuoka, M., Ishii, S., Shimizu, N., and Imaizumi, R. (1965). Effect of Win 18501-2 content of catecholamines and the number of catecholamine containing granules in the rabbit hypothalamus. *Experientia* 21:121–123.

Maynert, E. W. (1969). The role of biochemical and neurohumoral factors in the laboratory evaluation of antiepileptic drugs. *Epilepsia* **10**:145–162.

Maynert, E. W., and Levi, R. (1964). Stress-induced release of brain norepinephrine and its inhibition by drugs. *J. Pharm. Pharmacol.* **143**:90–95.

McCaman, R. E. (1965). Microdetermination of monoamine oxidase and 5-hydroxytryptophan decarboxylase activities in nervous tissues. *J. Neurochem.* **12**:15–23.

McCaman, R. E., and Aprison, M. H. (1964). The synthetic and catabolic enzyme systems for acetylcholine and serotonin in several discrete areas of the developing rabbit brain. *Progr. Brain Res.* **9**:220–223.

McCook, R. D., Peiss, C. N., and Randall, W. C. (1962). Hypothalamic temperatures and blood flow. *Proc. Soc. Exptl. Biol.* **109**:518–521.

McGeer, P. L. (1965). Subcellular localization of tyrosine hydroxylase in beef caudate nucleus. *Life Sci.* **4**:1859–1867.

McGeer, E. G., and McGeer, P. L. (1967). *In vitro* screen of inhibitors of rat brain tyrosine hydroxylase. *Can. J. Biochem.* **45**:115–131.

McGeer, P. L., McGeer, E. G., and Wada, J. A. (1963). Central aromatic amine levels and behavior. II. Serotonin and catecholamine levels in various cat brain areas following administration of psychoactive drugs or amine precursors. *Arch. Neurol.* **9**:81–89.

McIsaac, W. M., and Page, I. H. (1958). New metabolites of serotonin in carcinoid urine. *Science* **128**:537.

McIsaac, W. M., and Page, I. H. (1959). The metabolism of 5-HT. *J. Biol. Chem.* **234**:858–864.

McLennan, H. (1964). The release of acetylcholine and of 3-hydroxytyramine from the caudate nucleus. *J. Physiol. (Lond.)* **174**:152–161.

McLennan, H. (1965). The release of dopamine from the putamen. *Experientia* **12**:725–726.

McLennan, H. (1967). Evidence for synaptic transmitter functions. *Neurosci. Res. Prog. Bull.* **5**(1):64–66.

McLennan, H. (1969). *Synaptic Transmission,* W. B. Saunders Co., Philadelphia.

McNeill, J. H. (1964). The effects of phenelzine on serotonin, noradrenaline and monoamine oxidase in the rat. *Can. J. Physiol. Pharmacol.* **42**:33–39.

Mead, J. A. R., Shore, P. A., Weissbach, H., and Brodie, B. B. (1958). Comparative metabolism of brain serotonin and norepinephrine *in vitro*. *J. Pharm. Pharmacol.* **122**:51A.

Meltzer, H., Schander, R., and Grinspoon, L. (1969). The behavioral effects of nicotinamide adenine dinucleotide in chronic schizophrenia. *Psychopharmacolotia* **15**:144–151.

Mendelson, J., Kidzarisky, P., Lerderman, P. H., Wexler, D., Dutoit, C., and Solomon, P. (1960). Catecholamine excretion and behavior during sensory deprivation. *Arch. Gen. Psychiat.* **2**:147–155.

Menshikov, V. V. (1964). Spectorofluorometry of catecholamines with the use of a photoelectric apparatus for examination of combined dispersion spectra. *Vop. Med. Khim.* **10**:77–80.

Merrils, R. J. (1963). A semiautomatic method for determination of catecholamines. *Anal. Biochem.* **6**:272–282.

Messiha, F. S., Agallianos, D., and Clower, C. (1970). Dopamine excretion in affective states and following Li$_2$CO$_3$ therapy. *Nature* **225**:868–869.

Meyer, M. (1965). The effect of acetylcholine L-glutaminic acid and dopamine on neurons in the region of the nucleus cuneatus and nucleus gracilis in cats. *Helv. Physiol. Pharmacol. Acta* **23**:325–340.

Meyersson, B. J. (1964). Central nervous monoamines and hormone induced oestrus behavior in the spayed rat. *Acta. Physiol. Scand.* **63**: Suppl. 241.

Michaelson, I. A., and Whittaker, V. P. (1962). The distribution of 5-hydroxytryptamine in brain fractions. *Biochem. Pharmacol.* **11**:505.

Miller, M. L. (1934). Psychoses associated with probable injury to the hypothalamus and adjacent structures. *Arch. Neurol. Psychiat.* **31**:809.

Miller, R. E., Murphy, J. V., and Mirsky, I. A. (1957). Effect of chlorpromazine on fear motivated behavior of rats. *J. Pharmacol. Exptl. Therap.* **120**:379 *ff.*

Milline, R., Stern, P., and Hokovic, S. (1958). Sur les variations stressogénes quantitatives de la sérotonine dans le cerveau. *Experientia* **14**:418–426.

Milne, M. D., Crawford, M. A., Girac, C. B., and Longhridge, L. W. (1960). The metabolic disorder in Hartnup disease. *Quart. J. Med.* **29**:407–421.

Minz, B. (1957). Actions de drogues tranquillisantes sur le réaction du cortex cérébral à l'andrénaline. *Compt. Rend. Soc. Biol. (Paris)* **151**:432.

Minz, B., and Domino, E. (1952). Effects of epinephrine and norepinephrine on electrically induced seizures. *J. Pharmacol. Exptl. Therap.* **107**:204.

Minz, B., and Golstein, L. (1955). Epinephrine and cortical activity. *Fed. Proc.* **14**:371.

Minz, B., and Noel, P. (1961). Essay on cortico-hypothalamic-pharmacodynamics in the adrenalectomized rabbit. *Compt. Rend. Soc. Biol. (Paris)* **155**:1203–1207.

Minz, B., and Thuillier, J. (1959). Effects of iproniazid on hypothalamic reactions to adrenaline. *Compt. Rend. Soc. Biol. (Paris)* **153**:962–965.

Minz, B., and Walaszek, J. (1958). Topic action of epinephrine on the cerebral cortex of rabbits pretreated with serum from schizophrenics. *J. Pharmacol. Exptl. Therap.* **122**:53A.

Minz, B., and Walaszek, J. (1959). Effects of aminoxidase inhibitors on cerebral cortical responses to epinephrine. *Ann. N. Y. Acad. Sci.* **80**:617–626.

Minz, B., and Walaszek, J. (1960). Effects of serum of schizophrenics on epinephrine sensitive elements in the rabbit brain. *J. Nerv. Ment. Dis.* **130**:420–425.

Missala, K., Lloyd, K., and Gregoriads, G. (1967). Conversion of 14c-dopamine to cardiac 14c-noradrenaline in the copper-deficient rat. *Europ. J. Pharmacol.* **1**:6.

Mitro, A. (1965). Karyometric changes in the hypothalamus of the male albino rat during adaptation to repeated stress. I. NN. ventromediales, dorsomediales and arcuatus. *Biologia (Bratislava)* **20**:856–861.

Moguilevsky, J. A., Trifaro, J. M., Foglia, V. G., and Schiaffini, O. (1964). Effect of catecholamines on respiration of cerebral cortex and hypothalamus in female rats. *Am. J. Physiol.* **207**:733–735.

Moir, A. T. B. (1969). Effect of intravenous infusion of L-tryptophan on the cerebral metabolism of 5-hydroxyindoles and dopamine. *Biochem. J.* **114**:84P–85P.

Monnier, M., and Tissot, R. (1956). L'action de la réserpine-sérotonine sur le cerveau; suppression par les antagonistes de la réserpine: Iproniazide of LSD. *Arch. Suis. Neurol. Psychiat.* **82**:318–323.

Monnier, M., and Tissot, R. (1958). Action de la réserpine et de ces médiateurs (5HTP-sérotonine et DOPA-noradrenaline) sur les comportements et le cerveau du lapin. *Helv. Physiol. Acta* **16**:255–276.

Montagu, K. A. (1957). Catechol compounds in rat tissues and in brains of different animals. *Nature* **180**:244–245.

Montagu, K. A. (1963). Some catechol compounds other than noradrenaline and adrenaline in brains. *Biochem. J.* **86**:9–11.

Moore, K. E. (1966). Effects of alpha-methyltyrosine on brain catecholamines and conditioned behavior in guinea pigs. *Life Sci.* **5**:55–65.

Moore, K. E., and Brody, T. M. (1961). The effect of triethyltin on tissue amines. *J. Pharmacol. Exptl. Therap.* **132**:6–12.

Moore, K. E., and Lariviere, E. W. (1964). Effects of stress and D-amphetamine on rat brain catecholamines. *Biochem. Pharmacol.* **13**:1098–1100.

Moran, N. C. (1966). Pharmacological characterization of adrenergic receptors. *Pharmacol Rev.* **18**:503–512.

Morpurgo, C. (1962). Influence of phenothizine derivatives on the accumulation of brain amines induced by monoamine oxidase inhibitors. *Biochem. Pharmacol.* **11**:967–972.

Mouret, J. and Delorme, F. (1967). Lésions du tegmentum pontiqe et sommeil chez le rat. *C. R. Soc. Biol.* **161**:1603.

Murray, M. R. (1958). Response of oligodendrocytes to serotonin. In Windle, W. F. (ed.), *Biology of Neuroglia,* Charles C. Thomas, Springfield, Ill., pp. 176–190.

Musacchio, J., Kopin, I. J., and Snyder, S. (1964). Effects of disulfiram in tissue norepinephrine content and subcellular distribution of dopamine, tyramine and β-hydroxylated metabolites. *Life Sci.* 3:769–775.

Musacchio, J. M., Weise, V. K., and Kopin, I. J. (1965). Mechanism of norepinephrine binding. *Nature* 205:606–607.

Musacchio, J. M., Bhagat, B., Jackson, C. J., and Kopin, I. J. (1966a). The effect of disulfiram on the restoration of the response to tyramine by dopamine and α-methyl dopa in the reserpinized cat. *J. Pharmacol. Exptl. Therap.* 152:293–297.

Musaccio, J. M., Fischer, J. E., and Kopin, I. J. (1966b). Subcellular distribution and release by sympathetic nerve stimulation of dopamine and α-methyl dopamine. *J. Pharmacol. Exptl. Therap.* 152:51–55.

Musaccio, J. M., Goldstein, M., Anagoste, B., Poch, G., and Kopin, I. J. (1966c). Inhibition of dopamine β-hydroxylase by disulfiram *in vivo. J. Pharmacol. Exptl. Therap.* 152:56–61.

Muscholl, E. (1959). The action of nialamide on the concentration of catecholamines in heart and brain stem. *Med. Exp.* 1:363–367.

Muscholl, E. (1960). Die Hemmung der Noradrenalin-Aufnahme des Herzens durch Reserpin und die Wirkung von Tyramin. *Arch. Exptl. Pathol. Pharmakol.* 240:234–241.

Muscholl, E., and Vogt, M. (1957). The action of reserpine on sympathetic ganglia. *J. Physiol (Lond.)* 136:7P–8P.

Nagatsu, T. (1965a). Tyrosine hydroxylase, a specific enzyme involved in the biosynthesis of catecholamines in the brain and sympathetically innervated tissues. *Adv. Neurol. Sci.* 9:460–465.

Nagatsu, T. (1965b). Biosynthesis and metabolism of catecholamines. *J. Jap. Biochem. Soc.* 37:697–715.

Nagatsu, T., Levitt, M., and Udenfriend, S. (1964). Tyrosine hydroxylase; the initial step in norepinephrine biosynthesis. *J. Biol. Chem.* 239:2910–2917.

Nair, V. (1966). Regional changes in brain serotonin after head X-irradiation and its significance in the potentiation of barbiturate hypnosis. *Nature* 208:1293–1294.

Nakamura, S., Ichiyama, A., and Hayaishi, O. (1965). Purification and properties of tryptophan hydroxylase in brain. *Fed. Proc.* 24:604.

Nathan, H. A., and Friedman, W. (1962). Chlorpromazine affects the permeability of resting cells of tetrahymena pyriformis. *Science* 135:793–794.

Nicolosi, G., Santoro, R., Scarlata, S., and Ranzato, F. P. (1965). Il dosaggio delle catecolamine urinarie in soggetti schizofrenici prima e dopo elettroshock. *Acta Neurol. (Napoli)* 20:308–321.

Nielson, H. C. (1966). Effect of frontal pole ablation on biogenic amine levels in the brain. *Exptl. Neurol.* 15:484–489.

Norberg, K. A. (1965). Drug induced changes in monoamine levels in the sympathetic adrenergic ganglia cells and terminals. A histochemical study. *Acta Physiol. Scand.* 65:221–234.

Norberg, K. A., and Hamberger, B. (1964). The sympathetic adrenergic neurons. *Acta Physiol. Scand. (suppl.)* 63:238.

Norberg, K. A., and Sjöqvist, F. (1966). New possibilities for adrenergic modulation of ganglionic transmission. *Pharmacol. Rev.* 18:743–751.

Nukada, T. (1963). Monoamine oxidase activity and mitochondrial structure of the brain. *Jap. J. Pharmacol.* 13:124.

Olds, J. (1958). Self stimulation of the brain. *Science* 127:315–324.

Olds, J., and Olds, M. E. (1958). Positive reinforcement produced by stimulating the hypothalamus with iproniazid and other compounds. *Science* 127:1175–1176.

Olds, J., and Peretz, B. (1960). Motivational analysis of the reticular activating system. *Electroenceph. Clin. Neorophysiol.* 12:445–454.

Olds, J., Killam, K. F., and Eiduson, S. (1957). Effects of tranquilizers on self-stimulation of the brain. In Garattini, S., and Ghetti, V., (eds.), *Psychototropic Drugs*, Elsevier, Amsterdam, pp. 235–243.

Olds, J., Best, P. J., and Mink, W. (1967). Neuron correlates of motivational processes. *Science* **158**:533.

Ordy, J. M., Samorajski, T., and Schroeder, D. (1966). Concurrent changes in hypothalamic and cardiac catecholamine levels after anesthetics, tranquilizers and stress in a subhuman primate. *J. Pharmacol. Exptl. Therap.* **152**:445–457.

O'Reilly, S. (1965). The relationship between 5-hydroxyindoles and catecholamines. *Neurology* **15**:1142–1146.

Osmond, H., and Smythies, J. (1952). Schizophrenia: A new approach. *J. Ment. Sci.* **98**:309.

Osterholm, J. L., and Meyer, R. (1969). Experimental effects of free serotonin on the brain and its relation to brain injury. II: Trauma-induced alterations in spinal fluid and brain. *J. Neurosurg.* **31**:413–421.

Osterholm, J. L., Bell, J., Meyer, R., and Pyenson, R. (1969). Neurological consequences of intracerebral serotonin injections. *J. Neurosurg.* **31**:408–412.

Owman, C., and Rosengren, E. (1967). Dopamine formation in brain capillaries—An enzymic blood-barrier mechanism. *J. Neurochem.* **14**:547–550.

Paasonen, M. K., and Dews, P. B. (1958). Effects of raunescine and isoraunescine on behavior and on the 5-HT and noradrenaline contents of brain. *Brit. J. Pharmacol.* **13**:34–88.

Paasonen, M. K., and Giarman, N. J. (1958). Brain level of 5-HT after various agents. *Arch. Int. Pharmacodyn.* **114**:189–200.

Paasonen, M. K., and Vogt, M. (1956). The effects of drugs on the amount of substance P and 5-hydroxytryptamine in mammalian brain. *J. Physiol. (Lond.)* **131**:617–626.

Paasonen, M. K., McLean, P. D., and Giarman, N. J. (1957). 5-Hydroxytryptophamine (serotonin, enteramine) content of structures in the limbic system. *J. Neurochem.* **1**:326–333.

Palaic, D., and Supek, Z. (1966). Liberation of brain 5-hydroxytryptamine and noradrenaline by x-ray treatment in the new-born and adult rat. *J. Neurochem.* **13**:705–709.

Papez, J. W. (1958). Visceral brain, its component parts and their connections. *J. Nerv. Ment. Dis.* **126**:40.

Parkes, A. S., and Bruce, H. M. (1961). Olfactory stimuli in mammalian reproduction. *Science* **134**:1049.

Pazzagli, A., and Amaducci, L. (1966). La sperimentazione clinica del dopa nelle sindromi Parkinsoniane. *Riv. Neurobiol.* **12**:138.

Pellegrino de Iraldi, A. (1966). Noradrenaline and dopamine content of normal decentralized and denervated pineal gland of the rat. *Life Sci.* **5**:149–154.

Pellegrino de Iraldi, A., Farini-Duggau, H. J., and De Robertis, E. (1963). Adrenergic synaptic vesicles in the anterior hypothalamus of the rat brain. *Anat. Rec.* **145**:521–531.

Pennes, H. (1957). The nature of drugs with mental actions and their relation to cerebral function. *Bull. N. Y. Acad. Med.* **33**:81–88.

Perry, T. L., Hansen, S., and MacIntire, L. (1964). Failure to detect 3,4-dimethoxyphenylethylamine in urine of schizophrenics. *Nature* **202**:519–520.

Perry, T. L., Hansen, S., and MacDougall, L. (1967). Identity and significance of some pink spots in schizophrenia and other conditions. *Nature* **214**:484–485.

Persson, T., and Waldeck, B. (1968). The use of ^{3}H-dopa for studying cerebral catecholamine metabolism. *Acta Pharmacol. Toxicol.* **26**:363–372.

Petsche, H. (1966). Dopamingehalt im Caudatum des Kaninchens nach Pallidumläsionen. *Wien. Z. Nervenheilk.* **23**:117–121.

Peyrethon-Dusan, D. (1968). Etude dynamique et neuropharmacologique des phenomenes phasiques du sommeil paradoxal. Thesis. J. Tixier and Fils, Lyon, France, I-III.

Pfeiffer, C. C., Beck, R. A., and Neiss, E. S. (1967). Methylation or alkylation of biogenic amines as a factor in schizophrenias. In *Neuropsychopharmacology*, H. Brill (ed.), Excerpta Medica Foundation, New York, 1207.

Pfeiffer, A. K., Vizi, E. S., Satory, E., and Galambos, E. (1963). The effect of adrenalectomy on the norepinephrine and serotonin content of the brain and on the reserpine action in rats. *Experientia* **19**:482–483.

Philippu, A., and Schuemann, H. J. (1962). The influence of calcium on catecholamine liberation. *Experientia* **18**:138–140.

Philippu, A., and Shuemann, H. J. (1963). Effect of ribonuclease on the ribonucleic acid, adenosine triphosphate and catecholamine content of medullary granules. *Nature* **198**: 795–796.

Pick, P., and Feitelberg, S. (1948). Thermogenetic action of adrenaline and benzedrine on the brain. *Arch. Int. Pharmacodyn.* **77**:219–225.

Pickford, M. (1960). Factors affecting milk release in the dog and the quantity of oxytocin liberated by suckling. *J. Physiol. (Lond.)* **152**:515.

Piliego, N., and Rossini, P. (1963). Increase of catecholamines and vanillylmandelic acid in traumatic shock. *Boll. Soc. Ital. Biol. Sper.* **39**:603–606.

Pletscher, A. (1959). Significance of MAO inhibition for the pharmacological and clinical effects of hydrazine derivatives. *Ann N. Y. Acad. Sci.* **80**:1039–1045.

Pletscher, A. (1966). Monoamine oxidase inhibitors. *Pharmacol. Rev.* **18**:121–129.

Pletscher, A., and Besendorf, H. (1959). Antagonism between harmaline and long acting monoamine oxidase inhibitors concerning the effect on 5-HT and norepinephrine metabolism in brain. *Experientia* **15**:25–26.

Pletscher, A., and Da Prada, M. (1967). Mechanism of action of neuroleptics. In Brill, H. (ed.), *Neuropsychopharmacology,* Excerpta Medica Foundation, Amsterdam, pp. 304–311.

Pletscher, A., and Gey, K. F. (1962*a*). A new inhibitor of decarboxylase of aromatic amino acids. *Experientia* **18**:1–5.

Pletscher, A., and Gey, K. F. (1962*b*). Action of imipramine and amitriptyline on cerebral monoamines as compared with chlorpromazine. *Med. Exp.* **6**:165–168.

Pletscher, A., and Gey, K. F. (1962*c*). Interference with the permeation of aromatic monoamines and amino acids in brain, a possible new type of drug action. In *Proceedings of the First International Pharmacology Meeting,* Pergamon Press, London.

Pletscher, A., and Gey, K. F. (1963). Effect of a new decarboxylase inhibitor on endogenous and exogenous monoamines. *Biochem. Pharmacol.* **12**:223–228.

Pletscher, A., and Kunz, E. (1964). Accumulation of exogenous monoamines in brain *in vivo* and its alteration by drugs. *Progr. Brain Res.* **8**:45–52.

Pletscher, A., Shore, P. A., and Brodie, B. B. (1955). Serotonin release as a possible mechanism of reserpine action. *Science* **122**:374–375.

Pletscher, A., Shore, P. A., and Brodie, B. B. (1956). Serotonin as a mediator of reserpine action in brain. *J. Pharm. Pharmacol.* **116**:84–89.

Pletscher, A., Besendorf, H., and Gey, K. F. (1959). Depression of norepinephrine and 5-HT in the brain by benzoquinolizine derivatives. *Science* **129**:844.

Pletscher, A., Burkard, W. P., and Gey, K. F. (1964). Effect of monoamine releasers and decarboxylase inhibitors on the endogenous 5-hydroxyindole derivatives in brain. *Biochem. Pharmacol.* **13**:385–390.

Poirier, L. S., and Sourkes, T. L. (1965). Influence of the substantia nigra on the catecholamine content of the striatum. *Brain* **1**:181–192.

Porter, R. W. (1953). Hypothalamic involvement in the pituitary adrenocortical response to stress stimuli. *Am. J. Physiol.* **172**:515–519.

Poschel, B. P. (1965). Concerning the excitatory effect of methyl-metatyrosine on hypothalamic self-stimulation in the rat: A control study. *Life Sci.* **4**:53–56.

Poschel, B. P. (1966). Hypothalamic self stimulation: Its suppression by blockade of norepinephrine biosynthesis and reinstatement by metamphetamine. *Life Sci.* **5**:11–16.

Potter, L. T., and Axelrod, J. (1963). Properties of norepinephrine storage particles of rat heart. *J. Pharmacol. Exptl. Therap.* **142**:299–305.

Pscheidt, G. R. (1964). Monoamine oxidase inhibitors. *Internat. Rev. Neurobiol.* **7**:191–229.

Pscheidt, G. R. (1965). Chick brain amines: Normal levels and effect of reserpine and monoamine oxidase inhibitors. *Progr. Brain Res.* **16**:245–249.

Pscheidt, G. R., and Haber, B. (1965). Regional distribution of dihydroxyphenylalanine and 5-hydroxytryptophan decarboxylase and of biogenic amines in the chicken central nervous system. *J. Neurochem.* **12**:613–618.

Pscheidt, G. R., and Himwich, H. (1963*a*). Reserpine, monoamine oxidase inhibitors and distribution of biogenic amines in monkey brain. *Biochem. Pharmacol.* **12**:65–71.

Pscheidt, G. R., and Himwich, H. E. (1963*b*). Chicken brain amines with special reference to cerebellar norepinephrine. *Life Sci.* **7**:524–526.

Pujol, J. F. (1967). *Monoamines et sommeils. II. Aspects techniques et intérêt de l'étude du métabolisme central des monoamines au cours du sommeil,* J. Tixier & Fils, Lyon.

Purpura, D. P. (1956). Electrophysiological analysis of psychogenic drug action. I. Effect of LSD action on central synapses. *Arch. Neurol. Psychiat.* **75**:122–131/32–43.

Quay, W. B. (1963). Effect of dietary phenylalanine and tryptophan on pineal and hypothalamic serotonin levels. *Proc. Soc. Exptl. Biol.* **114**:718–721.

Quinn, G. P., Brodie, B. B., and Shore, P. A. (1958). Drug-induced release of norepinephrine in cat's brain. *J. Pharm. Pharmacol.* **122**:63A.

Ramirez, E., and Luza, S. (1967). Dimethyl sulfoxide in the treatment of mental patients. *Ann. N. Y. Acad. Sci.* **141**:655 *ff.*

Ramón y Cajal, S. (1903). Estudios talámicos. *Trab. Lab. Invest. (Madrid).*

Randrup, A., and Munkvad, I. (1969). Pharmacological studies on the brain mechanisms underlying two forms of behavioral excitation: Stereotyped hyperactivity and "rage." *Ann. N. Y. Acad. Sci.* **159**:928–938.

Rawson, R. O., and Hammel, H. T. (1963). Hypothalamic and tympanic temperatures in rhesus monkey. *Fed. Proc.* **22**:283.

Rebhun, J., Feinberg, S. M., and Zeller, E. A. (1954). Potentiating effect of iproniazid on action of some sympathomimetic amines. *Proc. Soc. Exptl. Biol.* **87**:218–220.

Rech, R. H., Borys, H. K., and Moore, K. E. (1966). Alterations in behavior and brain catecholamine levels in rats treated with α-methyltryrosine. *J. Pharmacol. Exptl. Therap.* **153**:412–419.

Redding, T. W. (1966). Effects of hypophysectomy on hypothalamic obesity in CBA mice. *Proc. Soc. Exptl. Biol.* **121**:726–729.

Reimert, H. (1963). Role and origin of noradrenaline in the superior cervical ganglion. *J. Physiol. (Lond.)* **167**:18–29.

Reinis, S. (1961). Changes in the relative vascularity of various parts of the central nervous system after the elimination of neocortex. *Acta Anat.* **46**:73–80.

Reis, D. J. (1965). Brain catecholamines: Relation to the defense reaction evoked by amygdaloid stimulation in cat. *Science.* **149**:450–451.

Reis, D. J., and Fuxe, K. (1969). Brain norepinephrine: Evidence that neuronal release is essential for sham rage behavior following brainstem transection in cat. *Proc. Natl. Acad. Sci.* **64**:108–112.

Reis, D. J., Miura, M., Weinbren, M., and Gunne, L. M. (1967). Brain catecholamines: Relation to defense reaction evoked by acute brainstem transection in cat. *Science* **156**:1768–1770.

Reis, D. J., Moorhead, D. T., and Merlino, N. (1970). Dopa-induced excitement in the cat. Its relationship to brain norepinephrine concentrations. *Arch. Neurol.* **22**:31–39.

Renault, J. (1967). *Monoamines et sommeils II. Role du système du raphé et de la sérotonine cérébrale dans l'endormissement,* J. Tixier & Fils, Lyon.

Resnick, O., Hagopian, M., Hoagland, H., and Freeman, H. (1960). An *in vivo* test for measuring MAO inhibition on human subjects. *Arch. Gen. Psychiat.* **2**:459–461.

Resnick, O., Krus, D. M., and Raskin, M. (1967). LSD-25–Drug interactions in man. In Brill, H. (ed.), *Neuropsychopharmacology,* Excerpta Medica Foundation, New York, p. 1103.

Resnick, O., Krus, D. M., and Raskin, M. (1968). LSD-25: Drug interactions in man. In Burger, A. (ed.), *Neuropsychopharmacology,* New York, Dekker.

Richardson, J. A., Woods, E. F., and Richardson, A. K. (1958). Elevation in plasma catecholamines following morphine. *J. Pharm. Pharmacol.* **122**:64A.

Rinaldi, F. (1958). The effects of two anti-serotonin substances BAS, BAB, and of bufotenin on the spontaneous activity of the brain. *J. Nerv. Ment. Dis.* **126**:272–283.

Rinne, J. K., Sonninen, V., and Palo, (1966). Excretion of homovanillic and vanillylmandelic acid in patients with extrapyramidal disorders. *Psychiat. Neurol. (Basel)* **151**:321–327.

Ritzén, M. (1966). Quantitative fluorescence microspectrophotometry of catecholamine-formaldenhyde products. *Exp. Cell Res.* **44**:505–520.

Rodriguez de Lores Arnaiz, G., and De Robertis, E. (1962). Cholinergic and non-cholinergic nerve endings in the rat brain. II. Subcellular localization of monoamine oxidase and succinate dehydrogenase. *J. Neurochem.* **9**:503–508.

Rogers, K. J., and Slater, P. (1971). Brain acetylcholine and monoamines during experimental catatonia. *J. Pharm. Pharmacol.* **23**:135–137.

Röhlich, P. (1965). Electron microscopy of the median eminence of the rat. *Acta Biol. Acad. Sci. Hung.* **15**:431–457.

Roos, B. E., and Werdinius, B. (1963). The effect of α-methyl DOPA on the metabolism of 5-hydroxytryptamine in brain. *Life Sci.* **2**:92–96.

Rosengren, A. M., and Rosengren, E. (1965). The occurrence of homovanillic acid in human brain. *Acta Pharmacol.* **23**:36–48.

Rossi, G. F., and Zanchetti, A. (1957). The brainstem reticular formation anatomy and physiology. *Arch. Ital. Biol.* **95**:199–435.

Rothballer, A. B. (1956). Studies of the adrenalin sensitive component of the reticular activating system. *Electroenceph. Clin. Neurophysiol.* **8**:603–621.

Rothballer, A. B. (1959). The effects of catecholamines on the central nervous system. *Pharmacol. Rev.* **11**:494–547.

Rothlin, E. (1957). Pharmacology of LSD and related compounds. In Garattini, S., and Ghetti, V. (eds.), *Psychotropic Drugs,* Elsevier, Amsterdam, p. 36.

Roussel, B. (1967). *Monoamines et sommeils IV. Suppression du sommeil paradoxale et diminution de la noradrenaline cerebrale par lesions des noyaux locus coeruleus,* J. Tixier & Fils, Lyon.

Ruckebusch, Y., Grivel, M. L., and Laplace, J. P. (1966). Behavioral and electrographic effects of the cerebroventricular injection of catecholamines in sleep. *Therapie* **21**:483–491.

Rudas, N. (1964). Influence of alcohol on the monoamines content of the central nervous system. *Acta Neurol. (Napoli)* **19**:848–849.

Ruthven, C. R. J., and Sandler, M. (1962). The estimation of homovanillic acid in urine. *Biochem. J.* **83**:30.

Salmoiraghi, G. C. (1966). Central adrenergic synapses. *Pharmacol. Rev.* **18**:717–726.

Salmoiraghi, G. C., and Bloom, F. E. (1964). Pharmacology of individual neurons. *Science* **144**:493–499.

Salmoiraghi, G. C., Bloom, F. E., and Costa, E. (1964). Adrenergic mechanisms in rabbit olfactory bulb. *J. Am. Physiol.* **207**:1417–1424.

Samorajski, T., and Marks, B. H. (1962). Localization of tritiated norepinephrine in mouse brain. *J. Histochem. Cytochem.* **10**:392–399.

Sano, I., Gamo, T., Kakimoto, Y., Taniguchi, K., Takesada, M., and Nishinuma, K. (1959). Distribution of catechol compounds in human brain, *Biochim. Biophys. Acta* **32**:584–587.

Sano, I., Taniguchi, K., Gamo, T., Takesada, M., and Kakimoto, Y. (1960). Die Katechinamine im Zentralnervensystem. *Klin. Wschr.* **38**:57–62.

Sano, K., Yoshioka, M., Ogashiwa, M., Ishijima, B., Ohye, C., Sekino, H., and Mayanagi, Y. (1967). Autonomic, somatomotor and electroencephalographic responses to stimulation of the human hypothalamus and the rostral brain stem. Third International Symposium on Stereoencephalatomy, Madrid, Spain.

Satinoff, E. (1967). Disruption of hibernation caused by hypothalamic lesions. *Science* **155**:1031–1033.

Satinoff, E. (1970). Hibernation and the central nervous system. *Progr. Physiol. Psychol.* **3**:201–236.

Schadé, J. P. (1959). Ionic and water movements across membranes of nerve cells in the cerebral cortex. *J. Neurophysiol.* **22**:245.

Schaepdryver, A., and Preziosi, P. (1959). Insulin and reserpin induced changes of the content of catecholamine ascorbic acid and cholestrol of the adrenal with or without iproniazid pretreatment. *Boll. Soc. Ital. Biol. Sper.* **35**:327–331.

Schanberg, S. M. (1963). A study of the transport of 5-hydroxytryptophan and 5-hydroxy-tryptamine (serotonin) into brain. *J. Pharmacol. Exptl. Therap.* **139**:191.

Schanberg, S. M., McIlroy, C. A., and Giarman, N. J. (1961). Uptake of 6-hydroxytryptophan by brain and its inhibition by certain aromatic amino acids. *Fed. Proc.* **20**:141.

Schanberg, S. M., Schildkraut, J. J., and Kopin, I. J. (1967). The effects of psychoactive drugs on norepinephrine-H^3 metabolism in brain. *Biochem. Pharmacol.* **16**:383.

Schayer, R. W. (1953). *In vivo* inhibition of monoamine oxidase studied with radioactive tyramine. *Proc. Soc. Exptl. Biol.* **84**:60–63.

Scheckel, C. L., and Boff, E. (1964). Behavioral effects of interacting imipramine and other drugs with *d*-amphetamine, cocaine and tetrabenazine. *Psychopharmacologia* **5**:198–208.

Scheckel, C. L., Boff, E., and Pazery, L. M. (1969). Hyperactive states related to the metabolism of norepinephrine and similar biochemicals. *Ann. N. Y. Acad. Sci.* **159**:939–950.

Schildkraut, J. J. (1965). The catecholamine hypothesis of affective disorders: A review of supporting evidence. *Am. J. Psychiat.* **122**:509–522.

Schildkraut, J. J., and Kety, S. S. (1967). Biogenic amines and emotion. *Science* **156**:21–30.

Schlesinger, K. (1965). Genetics of audiogenic seizures. I. Relation to brain serotonin and norepinephrine in mice. *Life Sci.* **4**:2345–2351.

Schmitt, H. (1963). Effect of catecholamines on the vasomotor centers. *Therapie* **18**:238–253.

Schmitt, H., and Schmitt, O. H. (1962). Modifications of the excitability of the hypothalamic vasopressor centers by local injections of catecholamines. *Experientia* **18**:565–566.

Schmitt, H., and Schmitt, O. H. (1964). Effets des injections locales et systémiques de catéchol-amines sur l'excitabilité des centres vasomoteurs diencephaliques et bulbaires. *Arch. Int. Pharmacodyn.* **150**:306–321.

Schneider, H. P. G., and McCann, S. M. (1970). Luteinizing hormone–releasing factor discharged by dopamine in rats. *J. Endocrinol.* **46**:401–402.

Schneider, J., and Thomalske, G. (1963). Untersuchungen über den Electrolythaushalt bei centrencephaler Epilepsie. *Nervenartz* **34**:338.

Schumann, H. J., and Kroneberg, G. (1970). *New Aspects of Storage and Release Mechanisms of Catecholamines,* Springer-Verlag, Berlin.

Schwartz, D. E., Burkard, W. P., Roth, M., Gey, K. F., and Pletscher, A. (1963). Effects of chlorpromazine on the penetration of monoamine oxidase inhibitors and monoamine releasers into rat brain. *Arch. Int. Pharmacodyn.* **141**:135–144.

Scott, J. W., and Pfaffman, C. (1967). Olfactory input to the hypothalamus: Electrophysiological evidence. *Science* **158**:1592–1594.

Segal, D. S., and Whalen, E. (1970). Effect of chronic administration of *p*-chlorophenylalanine on sexual receptivity of the female rat. *Psychopharmacologia* **16**:434–438.

Seiden, L. S. (1963). Mechanism of iproniazid inhibition of brain monoamine oxidase. *Arch. Int. Pharmacodyn.* **146**:145–162.

Sen, N. P., and McGeer, P. L. (1964). 4-Methoxyphenylethylamine and 3,4-dimethoxyphenyl-ethylamine in human urine. *Biochem. Biophys. Res. Commun.* **14**:227–232.

Sharp, J. C., Nielson, H. C., and Porter, P. B. (1962). The effect of amphetamine upon cats with lesions in the ventromedial hypothalamus. *J. Comp. Physiol. Psychol.* **55**:198–200.

Sheard, M. H., and Freedman, D. X. (1967). The effects of CNS lesions on norepinephrine, serotonin and acetylcholine in brain. *Brain Res.* **3**:292–294.

Sheppard, H., Tsien, W. H., Plummer, A. J., Peets, E. A., Giletti, B. U., and Schulert, A. R. (1958). Brain reserpine levels following large and small doses of reserpine H^3. *Proc. Soc. Exptl. Biol.* **97**:717–721.

Sherwood, S. L. (1955). Recent experiments with injections of drugs into the ventricular system of the brain. The response of psychotic patients to intraventricular injections. *Proc. Roy. Soc. Med.* **48**:855–863.

Shimuzu, N. (1964). Electron microscopic observation of catecholamine-containing granules in the hypothalamus and area postrema and their changes following reserpine injection. *Arch. Histol. Jap.* **24**:489–497.

Shimizu, N. (1965). Electron-miscrosopic observations on nucleolar extrusion in nerve cells of the rat hypothalamus. *Z. Zellforsch.* **67**:367–372.

Shimizu, N. (1966). On the catecholamine-containing granules in the brain. *Adv. Neurol. Sci.* **21**:4–15.

Shore, P. A. (1962). Release of serotonin and catecholamines by drugs. *Pharmacol. Rev.* **14**:531–550.

Shore, P. A., and Brodie, B. B. (1957). Influence of various drugs on serotonin and norepinephrine in brain. In Garattini, S., and Ghetti, V., (eds.), *Psychotropic Drugs*, Elsevier, Amsterdam, pp. 423–427.

Shore, P. A., and Olin, J. S. (1958). Identification and chemical assay of norepinephrine in brain. *J. Pharm. Pharmacol.* **122**:68A, 295–300.

Shore, P. A., Mead, J. A. R., Kuntzman, R. G., Spector, S., and Brodie, B. B. (1957a). On the physiological significance of monoamine oxidase in brain. *Science* **126**:1063–1064.

Shore, P. A., Pletscher, A., Tomich, E. G., Carlsson, A., Kuntzman, R., and Brodie, B. B. (1957b). Role of brain serotonin in reserpine action. *Ann. N. Y. Acad. Sci.* **66**:609–615.

Shore, P. A., Silver, S. L., and Brodie, B. B. (1957c). Interaction of reserpine, serotonin and LSD in brain. *Science* **122**:284–285.

Shute, C. C. D., and Lewis, P. R. (1966). Cholinergic and monoaminergic pathways. *Brit. Med. Bull.* **22**:222–226.

Siegfried, J., Klaiber, R., Perret, E., and Ziegler, W. H. (1969). The treatment of Parkinson's disease with L-dopa combined with a decarboxylase inhibitor. *Ninth Intern. Congr. Neurol.* New York, September 20–27.

Siegfried, J., Ziegler, W. H., Regli, F., Fischer, C., Kaufmann, N., and Perret, E. (1969). Treatment of Parkinsonism with L-dopa associated with a decarboxylase inhibitor. *Pharmacol. Clin.* **2**:23–26.

Sigg, E. B., Soffer, L., and Gyermek, J. (1963). Influence of imipramine and related psychoactive agents on the effect of 5-hydroxytryptamine and catecholamines on the cat nictitating membrane. *J. Pharmacol. Exptl. Therap.* **142**:13–20.

Simienesku-Karapancha, S., Trinkh-Binkh, D., and Babuianu, G. (1962). Studies on the relationship between serotonin and histamine in the mechanism of secretion of catecholamine. *Rev. Sci. Med. (Buc.)* **7**:119–122.

Sjöerdsma, A. (1959). Serotonin. *New Engl. J. Med.* **261**:181–188.

Sjöerdsma, A. (1966). Catecholamine drug interactions in man. *Pharmacol. Rev.* **18**:673–683.

Sjöerdsma, A., Smith, T. E., Stevenson, T. D., and Udenfriend, S. (1955). Metabolism of 5-HT by monoamine oxidase. *Proc. Soc. Exptl. Biol.* **89**:36–38.

Sjöerdsma, A., King, W. M., Leeper, L. C., and Udenfriend, S. (1958). Demonstration of the 3-methoxy analogue of norepinephrine in man. *Science* **127**:876.

Sjöerdsma, A., Gillespie, L., and Udenfriend, S. (1959). A method for measurement of MAO inhibition in man. *Ann. N. Y. Acad. Sci.* **20**:969–981.

Slimane, J., Taleb, S., and Torre, J. F. (1962). Morphological study of the paraventricular and the latero-hypothalamic accessory paraventricular nuclei in the male guinea pig after prolonged castration. *Compt. Rend. Soc. Biol. (Paris)* **156**:62–64.

Sloan, J. W., Brooks, J. W., Isenman, A. J., and Martin, W. R. (1962). Comparison of the effects of single doses of morphine and thebaine on body temperature, activity and brain and heart levels of catecholamines and serotonin. *Psychopharmacologia* **3**:291–301.

Smith, C. B. (1966). The role of monoamine oxidase in the intraneuronal metabolism of norepinephrine released by indirectly-acting sympathomimetic amines or by adrenergic nerve stimulation. *J. Pharmacol. Exptl. Therap.* **151**:2–7–220.

Smith, O. A., Jabbur, S. S., Rushmer, R. F., and Lasher, E. P. (1960). Role of hypothalamic functions in cardiac control. *Physiol. Rev. Suppl.* **4**:136–141.

Smith, P. E. (1931). Disorders induced by injury to the pituitary and the hypothalamus. *J. Nerv. Ment. Dis.* **74**:56.

Snyder, S. H., Fischer, J. E., and Axelrod, J. (1965). Evidence for presence of monoamine oxidase in sympathetic nerve endings. *Biochem. Pharmacol.* **14**:363–365.

Sokoloff, L., Perlin, S., Kornetsky, C., and Kety, S. S. (1965). The effects of *d*-lysergic acid dimethylamide on cerebral circulation and over all metabolism. *Ann. N. Y. Acad. Sci.* **66**:468–477.

Sommer, S. F., Novin, D., and LeVine, M. (1967). Food and water intake after intrahypothalamic injections of carbachol in the rabbit. *Science* **156**:983–984.

Soriani, S., and Favale, E. (1959). Protective action of LSD on behavior changes and on histological lesions of the C. N. S. induced in mice by chronic treatment with 5-HT. *Sist. Nerv.* **11**:16–21.

Soulairac, A., and Soulairac, A. M. (1964). Le role des catécholamines dans le contrôle de la neurosecrétion hypothalamique. *Agressologie* **5**:465–472.

Sourkes, T. L. (1958). Oxidative pathways in the metabolism of biogenic amines. *Rev. Can. Biol.* **17**:328–366.

Sourkes, T. L. (1963). Catecholamine metabolism and some functions of the nervous system. *Am. J. Clin. Nutr.* **12**:321–329.

Sourkes, T. L. (1966). Effect of brain stem lesions on the concentration of catecholamines in basal ganglia of the monkey. *J. Neurosurg.* **24**: Suppl. 194–195.

Sourkes, T. L., and Poirier, L. (1965). Influence of the substantia nigra on the concentration of 5-hydroxytryptamine and dopamine of the striatum. *Nature* **207**:202–203.

Sourkes, T. L., Drujan, B. D., and Curtis, G. C. (1959). Effect of repeated doses of insulin on excretion of pyrocatecholamines. *Arch. Gen. Psychiat.* **1**:265–268.

Sourkes, T. L., Murphy, G. F., Chavez, B., and Zilinska, M. (1961). The action of some α-methyl and other amino acids on cerebral catecholamines. *J. Neurochem.* **8**:109–115.

Sourkes, T. L., Murphy, G. F., Sankoff, I., Distler, M. H., and Saint Cyr, S. (1963). Excretion of dopamine, catecholamines, metabolites and 5-hydroxyindol-acetic acid in hepatolenticular degeneration (Wilson's disease) *J. Neurochem.* **10**:947.

Spector, S. (1963). Monoamine oxidase in control of brain serotonin and norepinephrine content. *Ann. N. Y. Acad. Sci.* **107**:856–861.

Spector, S., Prockop, D., Shore, P. A., and Brodie, B. B. (1958). The effect of iproniazid on brain levels of norepinephrine and serotonin. *Science* **127**:704.

Spector, S., Shore, P. A., and Brodie, B. B. (1960). Biochemical and pharmacological effects of the MAO inhibitors, iproniazid, 1-phenyl-2-hydrazinepropane (JB 516) and 1-phenyl-3-hydrazinebutane (JB 835). *J. Pharmacol. Exptl. Therap.* **128**:15–21.

Spector, S., Melmon, K., and Sjöerdsma, A. (1962). Evidence for rapid turnover of norepinephrine in rat heart and brain. *Proc. Soc. Exptl. Biol.* **111**:79–81.

Spector, S., Sjöerdsma, A., and Udenfriend, S. (1965). Blockade of endogenous norepinephrine synthesis by -methyltyrosine, an inhibitor of tyrosine hydroxylase. *J. Pharmacol. Exptl. Therap.* **147**:86–95.

Spilman, E. L., and Badal, D. W. (1960). Effect of electroconvulsive therapy on brain amino oxidase. *Arch. Gen. Psychiat.* **2**:545–547.

Stark, P., Boyd, E. E., and Fuller, R. W. (1964). A possible role of serotonin in hypothalamic self-stimulation in dogs. *J. Pharmacol. Exptl. Therap.* **146**:147–153.

Starykh, N. T. (1962). Effect of amizil on the content of serotonin in some segments of the cat brain. *Biull. Eksp. Biol. Med.* **54**:76–78.

Steg, G. (1969). Side-effects during treatment with L-dopa in Parkinsonism. Ninth International Congress of Neurology, New York, September 20–27, 1969.

Stein, L. (1968). Chemistry of reward and punishment. In Efron, D. H. (ed.), *Psychopharmacology: A Review of Progress 1957–1967*. Public Health Service Publication No. 1836, 105–123.

Stein, L., and Seifter, J. (1962). Muscarinic synapses in the hypothalamus. *Am. J. Physiol.* **202**: 751–756.

Stein, L., and Wise, C. D. (1971). Possible etiology of schizophrenia: Progressive damage to the noradrenergic reward system by 6-hydroxy dopamine. *Science* **171**:1032–1036.

Stjärne, L. (1964). Studies of catecholamine uptake, storage and release mechanisms. *Acta Physiol. Scand.* **62**: (Suppl. 228) 1–77.

Stjärne, L. (1966). Recent observations on the regulation of storage and release mechanisms. *Res. Publ. Ass. Res. Nerv. Ment. Dis.* **43**:325–342.

Stone, C. A., Ross, C. A., Wenger, H. C., Ludden, C. T., Blessing, J. A., Totaro, J. A., and Porter, C. C. (1962). Effect of α-methyl-3,4-dihydroxyphenylalanine (α-methyl dopa), reserpine and related agents on some vascular responses in the dog. *J. Pharm. Pharmacol.* **136**:80–88.

Stoupel, N., and Terzuolo, C. (1954). Etude éléctrophysiologique des connexions et de la physiologie du noyau caudé. *Acta Neurol. Belg.* **54**:239–248.

Ström, G. (1950). Influence of local thermal stimulation of the hypothalamus of the cat on cutaneous blood flow and respiratory rate. *Acta Physiol. Scand.* **20**: (Suppl. 70) 47–76.

Strom-Olsen, R., and Weil-Malherbe, H. (1958). Humoral changes in manic-depressive psychoses with particular reference to the excretion of catecholamines in urine. *J. Ment. Sci.* **104**:696–704.

Strubelt, O. (1964). The effect of ionizing radiations on the monoamine oxidase activity in liver and brain of white mice. *Strahlentherapie* **124**:570–572.

Sudak, H. S., and Maas, J. W. (1964). Central nervous system serotonin and norepinephrine localization in emotional and non-emotional strains in mice. *Nature* **203**:1254–1256.

Sudakov, K. V. (1965). On the interaction of hypothalamus, reticular formation of the mesencephalon and thalamus in the mechanism of selective ascending activation of the cerebral cortex during physiologic hunger. *Fiziol. Zh. (Mosk.)* **51**:449–456.

Sulser, F., and Brodie, B. B. (1960). Is reserpine tranquilization linked to change in brain serotonin or brain epinephrine? *Nature* **131**:1440–1441.

Sulser, F., Bickel, M. H., and Brodie, B. B. (1964). The action of desmethylimipramine in counteracting sedation and cholinergic effects of reserpine-like drugs. *J. Pharmacol. Exptl. Therap.* **144**:321–330.

Sunderman, F. W. (1963). Disturbances of indole metabolism in hepatolenticular degeneration. *Am. J. Med. Sci.* **246**:165–171.

Szara, S. (1964). Behavioral correlates of 6-hydroxylation and the effect of psychotropic tryptamine derivatives on brain serotonin levels. In Richter, D. (ed.), *Comparative Neurochemistry,* Pergamon Press, Oxford, pp. 425–432.

Szara, S. (1968). Discussion of the fate and metabolism of some hallucinogenic indolealkylamines. *Adv. Pharmacol.* **6B**:230–231.

Szentagothai, J., Flerko, B., Mess, B., and Halasz, B. (1962). *Hypothalamic Control of the Anterior Pituitary,* Academiai Kiado, Budapest.

Tadachnick, I. I., and Rubin, A. A. (1959). Some relationships between peripheral monoamine oxidase inhibition and brain 5-hydroxytryptamine levels in rats. *Proc. Soc. Exptl. Biol.* **101**:435–437.

Taeschler, M. (1965). Pharmacological and clinical action of antidepressive drugs. Pharmacological aspects. In Bente, D., and Bradley, P. B. (eds.), *Neuropsychopharmacology,* Elsevier, Amsterdam, Vol. IV, pp. 197–202.

Taeschler, M. (1967). Pharmacology of psychotomimetic agents. In Brill, H. (ed.), *Neuropsychopharmacology,* Excerpta Medica Foundation, Amsterdam, pp. 393–397.

Tagliamonte, A., Tagliamonte, P., Gessa, G. L., and Brodie, B. B. (1969). Compulsive sexual activity induced by *p*-chlorophenylalanine in normal and pinealectomized male rats. *Science* **166**:1433–1434.

Takashima, H. (1962). Relation between pyrexia and amine levels in brain with special reference to the action of monoamine oxidase inhibitor. *Folia Pharmacol. Jap.* **58**:437–448.

Takesade, M., Kakimoto, Y., and Kaneko, Z. (1963). 3,4-Dimethoxyphenylethylamine and other amines in the urine of schizophrenic patients. *Nature* **199**:203–204.

Taniguchi, K., Kakimoto, Y., and Armstrong, M. D. (1964). Quantitative determination of metanephrine and normetanephrine in urine. *J. Lab. Clin. Med.* **64**:469–484.

Terranova, R., Vanni, F., and Spiazzi, R. (1962). Permeability of neuronal membrane to psychotropic drugs. In *Fourth International Congr. of Neuropathology, Monaco*, G. Thieme Verlag, Stuttgart, p. 191.

Thieme, G. (1965). Small tissue dryers with high capacity for rapid freeze-drying. *J. Histochem. Cytochem.* **13**:386–389.

Thoenen, H., and Tranzer, J. P. (1968). Chemical sympathectomy by selective destruction of adrenergic nerve endings with 6-hydroxy dopamine. *Arch. Pharmakol. Exptl. Pathol.* **261**:271–288.

Timiras, P. S., Woodbury, D. M., and Goodman, L. S. (1954). Effect of adrenalectomy, hydrocortisone acetate on brain excitability and electrolyte distribution in mice. *J. Pharmacol. Exptl. Therap.* **110**:80–93.

Tissot, R. (1961). Monoamines et système nerveux central. *Encephale* **40**:105–179, 205–306.

Tissot, R., and Monnier, M. (1958). Suppression de l'action de la réserpine sur le cerveau par ces antagonistes: Iproniazid et LSD. *Helv. Physiol. Acta* **16**:268–276.

Tissot, R., Guggisberg, M., and Constantinidis, J. (1966). Excretion urinaire de l'acide 5-hydroxyindolacétique, de l'acide vanillomandelique, et de l'acide xanthurénique chez 18 mongoliens. *Pathol. Biol.* **14**:312–316.

Tissot, R., Gaillard, J. M., Guggisberg, M., Gauthier, G., and Ajuriaguerra, J. de (1969). Therapeutique du syndrome de Parkinson par la dopa "per os" associée à un inhibiteur de la decarboxylase (Ro 4-4602). *Presse Méd.* **77**:619–622.

Titus, E., and Udenfriend, S. (1954). Metabolism of 5-hydroxytryptamine (serotonin). *Fed. Proc.* **13**:411.

Tonkikh, A. V. (1965). Changes in the electric activity of the hypothalamus following excitation of the sensory nerve and epinephrine administration. *Fizio. Zh. (Mosk.)* **51**:755–761.

Tower, D. B. (1960). *Neurochemistry of Epilepsy*, Charles C. Thomas, Springfield, Ill., p. 188.

Towne, J. C. (1964). Effect of ethanol and acetaldehyde on liver and brain monoamine oxidase. *Nature* **201**:709–710.

Townsend, E. E., Katis, J. G., and Sourkes, T. L. (1958). Effect of tryptophan deficiency in rat on liver tryptophan peroxidase (TPO) and urinary 5-hydroxy-3-indoleacetic acid (5-HIAA). *Fed. Proc.* **17**:324.

Tranzer, J. P., and Thoenen, H. (1968). An electron microscopic study of selective, acute degeneration of sympathetic nerve terminals after administration of 6-hydroxydopamine. *Experientia* **24**:155–156.

Triped, J., Studer, A., and Meier, R. (1957). Essai de différenciation expérimentale d'une série d'inhibiteurs du système nerveux central. *Arch. Int. Pharmacodyn.* **112**:319.

Trzabski, A. (1962). Studies on the central effect of neurohormones. II. Effect of serotonin and reserpine on cardiovascular centers of the hypothalamus and on the reticular formation of the brain stem. *Acta Physiol. Pol.* **13**:77–91.

Udenfriend, S., and Creveling, C. R. (1959). Localization of dopamine-beta-oxidase in brain. *J. Neurochem.* **4**:350–352.

Udenfriend, S., and Weissbach, H. (1958). Turnover of 5-HT in tissues. *Proc. Soc. Exptl. Biol.* **97**:7848–7851.

Udenfriend, S., and Zaltzman-Nirenberg, P. (1963). Norepinephrine and 3,4-dihydroxyphenylethylamine turnover in guinea pig brain *in vivo. Science* **142**:394–396.

Udenfriend, S., Clark, C. T., and Titus, E. (1953). Hydroxylation of the 5-position of tryptophan as first step in its metabolic conversion to 5-hydroxytryptamine (serotonin). *Fed. Proc.* **12**:282.

Udenfriend, S., Clark, C. T., Axelrod, J., and Brodie, B. B. (1954). Ascorbic acid in aromatic hydroxylation. *J. Biol. Chem.* **208**:731.

Udenfriend, S., Weissbach, H., and Clark, C. T. (1955). The estimation of 5-HT in biological tissues. *J. Biol. Chem.* **215**:337–344.

Udenfriend, S., Titus, E., Weissbach, H., and Peterson, R. E. (1956). Biogensis and metabolism of 5-hydroxyindole compounds. *J. Biol. Chem.* **219**:335–344.

Udenfriend, S., Weissbach, H., and Bogdanski, D. E. (1957*a*). Increase in tissue serotonin following administration of its precursor 5-hydroxytryptophan. *J. Biol. Chem.* **224**:803–810.

Udenfriend, S., Weissbach, H., and Bogdanski, D. E. (1957*b*). Biochemical studies on serotonin and their physiological implications. In *Hormones, Brain Function and Behavior*, Academic Press, New York, pp. 147–160.

Udenfriend, S., Lovenberg, W., and Weissbach, H. (1960). L-amino acid decarboxylase activity in mammalian tissues and its inhibition by α-methyl-dopa. *Fed. Proc.* **19**:7.

Udenfriend, S., Zaltzman-Nirenberg, P., and Nagatsu, T. (1965). Inhibitors of purified beef adrenal tyrosine hydroxylase. *Biochem. Pharmacol.* **14**:837–845.

Umbach, W., and Tzavellas, O. (1965). Behandlung akinetischer Begleitsymptome beim Parkinson-Sydrom. *Deutsch. Med. Wschr.* **90**:1941–1944.

Ungerstedt, U., Butcher, L. L., and Butcher, S. G. (1969). Direct chemical stimulation of dopaminergic mechanisms in the neostinatum of the rat. *Brain Res.* **14**:461.

Uretsky, N. J., and Iversen, L. L. (1970). Effects of 6-hydroxydopamine on catecholamine containing neurons in the rat brain. *J. Neurochem.* **17**:269–278.

Utley, J. C. (1965). Relative effects of L-DOPA and its methyl ester given orally or intraperitoneally to reserpine-treated mice. *Acta Pharmacol.* **23**:189–193.

Utley, J. C. (1966). Effects of anthranil- benzo-, and salicyl-hydroxamic acids on mouse brain amines and L-dopa decarboxylase. *Proc. Soc. Exptl. Biol.* **123**:131–132.

Van Orden, L. S., III. (1970). Quantitative histochemistry of biogenic amines. A simple microspectrofluorometer. *Biochem. Pharmacol.* **19**:1105–1117.

Vernikos-Danellis, J. (1965). Effect of stress, adrenalectomy, hypophysectomy and hydrocortisone on the cortisotropin-releasing activity of rat median eminence. *Endrocrinology* **76**:122.

Verster, J. P. (1963). Preliminary report on the treatment of mentally disordered patients by intrathecally administered phenothiazine drugs and an anti-serotonin substance. *S. Afr. Med. J.* **37**:1068–1089.

Viteck, V. (1964). Biochemical aspects of catecholamines in the physiology of the central nervous system. *Wien. Z. Nervenheilk.* **22**:138–161.

Vogt, M. (1954). The concentration of sympathin in different parts of the central nervous system under normal conditions and after the administration of drugs. *J. Physiol. (Lond.)* **123**:541.

Vogt, M. (1959). Catecholamines in brain. *Pharmacol. Rev.* **11**:483–489.

Vogt, M. (1965). Effect of drugs on metabolism of catecholamines in the brain. *Brit. Med. Bull.* **21**:57–61.

Vogt, M., Gunn, C. G., and Sawyer, C. H. (1957). Electroencephalographic effects of intraventricular 5-HT and LSD in cat. *Neurology* **7**:559–566.

Vyshnepol'skii, Y. Y. (1966). Value of investigation hourly excretion of catecholamines in various portions of urine collected and following hypertensive crisis. *Fed. Proc.* **25**: (Suppl.) 471–472.

Wada, J. A. (1961). Epileptogenic cerebral electrical activity and serotonin levels. *Science* **134**:1688–1690.

Wada, J. A. (1966). Central aromatic amines and behavior. (3) Correlative analysis of conditioned approach behavior and brain levels of serotonin and catecholamines in monkeys. *Arch. Neurol.* **14**:129–142.

Wada, J. A., Wrinch, J., Hill, D., McGeer, P. L., and McGeer, E. G. (1963). Central aromatic amine levels and behavior. I. Conditioned avoidance response in cats, following administration of psychoactive drugs or amine percursors. *Arch. Neurol.* **9**:69–80.

Walaszek, E., and Abood, L. (1959). Fixation of 5-hydroxytryptamine by brain mitochondria, *Proc. Soc. Exptl. Biol.* **101**:37–40.

Walaszek, E., Smith, C. M., and Minz, B. (1958). Effect of serum from schizophrenic patients on neurohormones in rabbit brain. *Fed. Proc.* **17**:416.

Wase, A. W., Christensen, J., and Pelley, E. (1956). The accumulation of S-35 chlorpromazine in brain. *Arch. Neurol. Psychiat.* **75**:54–56.

Wawrzyniak, M. (1965). The histochemical activity of some enzymes in the mesencephalon during ontogenic development of the rabbit and guinea pig. II. Histochemical development of acetylcholinesterase and monoamine oxidase in the nontectal portion of the midbrain of the guinea pig. *Z. Mikr. Anat. Forsch.* **73**:261–305.

Weber, L. J. (1966). Influence of monoamine oxidase inhibitors on 5-hydroxytryptamine synthesis in the brain. *Proc. Soc. Exptl. Biol.* **123**:35–38.

Weight, F., and Salmoiraghi, G. C. (1965). Responses of single spinal cord neurons to ACh, NE and 5-HT administered by micro-electrophoresis. *Pharmacologist* **7**:174.

Weight, F., and Salmoiraghi, G. C. (1966). Responses of spinal cord interneurons to acetylcholine, norepinephrine and serotonin administered by microelectrophoresis. *J. Pharmacol. Exptl. Therap.* **153**:420–427.

Weight, F., and Salmoiraghi, G. C. (1967). Motoneurone depression by norepinephrine. *Nature* **213**:1229–1230.

Weil-Malherbe, H., and Bone, A. D. (1957). Intracellular distribution of catecholamines in the brain. *Nature* **180**:1050–1051.

Weil-Malherbe, H., Axelrod, J., and Tomchick, R. (1959). Blood–brain barrier for adrenalin. *Science* **129**:1226–1227.

Weil-Malherbe, H., Posner, H. S., and Bowles, G. P. (1961*a*). Changes in the concentration and intracellular distribution of brain catecholamines: The effects of reserpine, β-phenylisopropylhydrazine, pyrogallol, and 3,4-dihydroxyphenylalanine alone and in combination. *J. Pharmacol. Exptl. Therap.* **132**:272–286.

Weil-Malherbe, H., Whitby, L. G., and Axelrod, J. (1961*b*). The blood–brain barrier for catecholamines. In *Regional Neurochemistry,* Pergamon Press, New York, pp. 284–292.

Weinshilboum, R., and Axelrod, J. (1971). Serum dopamine-β-hydroxylase activity. *Circ. Res.* **28**:307–315.

Weissman, A., and Finger, K. F. (1962). Effects of benzquinamide on avoidance behavior and brain amine levels. *Biochem. Pharmacol.* **871**:80.

Welsh, J. H., and Moorehead, M. (1960). The quantitative distrubition of 5-hydroxytryptamine in the invertebrates specially in their nervous systems. *J. Neurochem.* **6**:146.

Welch, B. L., and Welch, A. M. (1964). An effect of aggregation upon the metabolism of dopamine-1-^{3}H. *Progr. Brain Res.* **8**:201–206.

Whitby, L. G., Axelrod, J., and Weil-Malherbe, H. (1961). Fate of H^3-norepinephrine in animals. *J. Pharmacol. Exptl. Therap.* **132**:193–201.

Whittaker, V. P. (1959). A comparison of the distribution of lysosome enzymes and 5-hydroxytryptamine with that of acetylcholine in subcellular fractions of guinea pig brain. *Biochem. Pharmacol.* **1**:351.

Whittaker, V. P. (1966). Catecholamine storage particles in the central nervous system. *Pharmacol. Rev.* **18**:401–412.

Wilson, C. W., Murray, A. W., and Titus, E. (1962). The effects of reserpine on the uptake of epinephrine in brain and certain areas outside the blood–brain barrier. *J. Pharmacol. Exptl. Therap.* **135**:11–16.

Wolfe, D., Potter, L. T., Richardson, K., and Axelrod, J. (1962). Localizing tritiated norepinephrine in sympathetic axons by electron microscopy and auto-radiography. *Science* **138**:440.

Wood, J. G. (1966*a*). Electron-microscopic localization of 5-hydroxytryptamine (5-HT). *Texas Rep. Biol. Med.* **23**:828–837.

Wood, J. G., (1966*b*). Electron localization of amines in central nervous tissue. *Nature* **209**: 1131–1133.

Woodford, V. R., and Barthwal, J. P. (1964). The effect of dietary deficiencies of tryptophan and niacin on catecholamine production in the rat. *Can. J. Biochem.* **42**:889–896.

Woolley, D. W. (1958). Neurologic and psychiatric changes related to serotonin. *Progr. Neurobiol.* **3**:157–170.

Woolley, D. W. (1962). *The Biochemical Basis of Psychosis,* John Wiley and Sons, Inc., New York.

Woolley, D. W., and Shaw, E. (1954). Biochemical and pharmacological suggestion about certain mental disorders. *Proc. Natl. Acad. Sci.* **40**:228.

Woolley, D. W., and Shaw, E. (1957). Evidence for the participation of serotonin in mental processes. *Ann. N. Y. Acad. Sci.* **66**:649–667.

Wurtman, R. J., and Axelrod, J. (1963). Sex steroids, cardiac ^{3}H-norepinephrine and tissue monamine oxidase levels in rat. *Biochem. Pharmacol.* **12**:1417–1419.

Wyatt, R. J., Engelman, K., Kupfer, D. J., Scott, J., Sjöerdsma, A., and Snyder, F. (1969). Effects of *p*-chlorophenylalanine on sleep in man. *Electroenceph. Clin. Neurophysiol.* **27**:529–532.

Yahr, M. D., Duvoisin, R. C., Mendoza, M. R., Shear, M. J., and Barrett, R. E. (1972). Modification of L-dopa therapy of Parkinsonism by alpha-methyl dopa hydrazine (MK-486). *Trans. Am. Neurol. Assoc.* (in press).

Yokohama, H. (1964). Catecholamine release by beta-phenyl-ethylamine, tyramine and sympathetic ganglionic stimulating drugs. *Folia Pharmacol. Jap.* **60**:293–307.

Yuwiler, A., Geller, E., and Slater, G. G. (1965). On the mechanism of the brain serotonin depletion in experimental phenylketonuria. *J. Biol. Chem.* **240**:1170–1174.

Zeller, E. A. (1959). The role of amine oxidases in the destruction of catecholamines. *Pharmacol. Rev.* **11**:387 *ff*.

Zeller, E. A. (1963). A new approach to the analysis of the interaction between monoamine oxidase and its substrates and inhibitors. *Ann N. Y. Acad. Sci.* **107**:811–821.

Zeller, E. A., and Barsky, J. (1952). *In vivo* inhibition of liver and brain monoamine oxidase by 1-isonicotinyl-2-isopropylhydrazine. *Proc. Soc. Exp. Biol. Med.* **81**:459.

Zeller, E. A., Barsky, J., and Berman, E. R. (1955). Amineoxidases: Inhibition of monoamine oxidase isonicotinyl-2-isopropylhydrazine. *J. Biol. Chem.* **214**:267–273.

Zeman, W., and Innes, J. R. M. (1963). *Craigie's Neuroanatomy of the Rat,* Academic Press, New York.

Zhukova, T. P. (1961). Multiplication of capillaries in different parts of the brain in the postnatal period. *Biull. Eksptl. Biol. Med.* **51**:87–93.

Index